Morgan George Watkins

Gleanings from the Natural History of the Ancients

Morgan George Watkins

Gleanings from the Natural History of the Ancients

ISBN/EAN: 9783337026486

Printed in Europe, USA, Canada, Australia, Japan

Cover: Foto ©berggeist007 / pixelio.de

More available books at **www.hansebooks.com**

GLEANINGS

FROM THE

NATURAL HISTORY OF THE ANCIENTS.

GLEANINGS

FROM THE

NATURAL HISTORY OF THE ANCIENTS.

BY

REV. M. G. WATKINS, M.A.

'Αλλ' οὐ μέντοι σοι, ἦν δ' ἐγώ, 'Αλκίνου γε ἀπόλογον ἐρῶ.
Platonis Republica, x. 614.

LONDON: ELLIOT STOCK,
62, PATERNOSTER ROW, E.C.
1885.

CONTENTS.

CHAPTER						PAGE
INTRODUCTION	-	-	-	-	-	vii–xiii
I. A HOMERIC BESTIARY	-	-	-	-	-	1
II. GREEK AND ROMAN DOGS	-	-	-	-	-	20
III. ANTIQUARIAN NOTES ON THE BRITISH DOG	-	-	-	-	-	31
IV. THE CAT	-	-	-	-	-	53
V. OWLS	-	-	-	-	-	68
VI. PYGMIES	-	-	-	-	-	80
VII. ELEPHANTS	-	-	-	-	-	88
VIII. THE HORSE	-	-	-	-	-	103
IX. GARDENS	-	-	-	-	-	125
X. HUNTING AMONG THE ANCIENTS	-	-	-	-	-	142
XI. THE ROMANS AS ACCLIMATIZERS IN BRITAIN	-	-	-	-	-	156
XII. VIRGIL AS AN ORNITHOLOGIST	-	-	-	-	-	172
XIII. ROSES	-	-	-	-	-	189
XIV. WOLVES	-	-	-	-	-	201
XV. ANCIENT FISH-LORE	-	-	-	-	-	213
XVI. MYTHICAL ANIMALS	-	-	-	-	-	228
XVII. OYSTERS AND PEARLS	-	-	-	-	-	244

INTRODUCTION.

THESE chapters, on a few of the curiofities connected with the natural
hiftory of the ancients, are in fome
refpects a faithful reflection of that
knowledge. They are fragmentary, and greatly
indebted to the labours of previous workers. But
they have not been put together without much
trouble and not a little honeft, diligent refearch ;
my object being to collect fome of the more interefting facts bearing upon ten or a dozen different
fubjects, rather than to write a complete natural
hiftory of the ancients. I have generally traced
thefe curious beliefs through their mediæval modi-
fications ; partly that the reader might be led to
contraft them with the exacter knowledge of the
prefent day, partly in order to fhew their growth
from, in fome cafes, pre-hiftoric and geological
times.

No one is more aware of the incompletenefs of
thefe Effays, yet I venture to hope that fome may

find in reading them a little of the fame pleafure
which I have experienced while fearching for the
facts they contain among the lefs frequently ex-
plored by-paths of claffical literature. They are, at
all events, a contribution to a fafcinating ftudy—
fpeculations rendered venerable by their antiquity,
rather than by the credit due to the writers who
are here laid under contribution. I would fain
fhelter, therefore, under Lord Bacon's mantle:
"Summæ pufillanimitatis eft auctoribus infinita
tribuere, auctori autem auctorum, atque adeo
omnis auctoritatis, tempori, jus fuum denegare.
Recte enim veritas temporis filia dicitur non
auctoritatis."[1]

He who has been accuftomed to teft modern
biological problems by means of the inductive
philofophy, is ftruck with amazement when he
firft turns to the natural hiftory of the ancients.
There are many regular writers of it; many
fcattered allufions to and accounts of animal life
in the poets. But all the natural hiftory of the
ancients labours under the fame faults, faults
infeparable, however, from the infancy of the
race—an inability to difcriminate with any accuracy, great ignorance of anatomy and phyfiology,
and a habit of accepting ftatements on infufficient
evidence. The writers of ancient natural hiftory
were, to ufe a modern phrafe of pregnant meaning,
wholly uncritical. Poetry and folk-lore were
confufed with exact fcience. Like children, they
were quick to grafp at marvels, to embrace a

[1] "Nov. Organum," i. 84.

narrative eagerly the more marvellous that it was.
Anything in the nature of a traveller's ſtory they
welcomed as readily as we ſhould diſtruſt it.
"Thus the crocodile from an egg growing up to
an exceeding magnitude, common conceit and
divers writers deliver, it hath no period of en-
creaſe, but groweth as long as it liveth. And thus,
in brief, in moſt apprehenſions the conceits of
men extend the conſiderations of things, and
dilate their notions beyond the propriety of their
natures."—(Sir T. Browne, "Vulgar Errors,"
vii. 15.)

It has often been queſtioned whether Herodotus was really impoſed upon by the Egyptian
prieſts or not. In either caſe the reſult, ſo far as
he is concerned, is the ſame. Many of the
marvels in the "Odyſſey" are exaggerations and
diſtortions of merchants' and ſailors' narratives.
While they accepted all that was told them without much queſtioning or heſitation, the ancient
writers of natural hiſtory never dreamt of teſting
any concluſion by obſervation, much more by ex-
periment. Pliny relates a thouſand marvels which
he might have omitted or modified had he taken
the trouble to conſult nature. But a naturaliſt,
in his acceptation of the term, meant little more
than a compiler and tranſcriber. From this miſtaken view, natural hiſtorians among the ancients
were quick to follow previous writers, and it is
not ſurpriſing to find blunders and miſconceptions
thus repeated over and over again. No muſeums
or collections enabled them to correct wrong im-

preffions. Later hiftorians were willing to believe
the marvels fet forth by their predeceffors, and
fo long as they did not deem it a part of their
duty to make original inquiries, it was inevitable
that hippogryphs, harpies, chimæras, and many
more fabulous monfters were handed on from
generation to generation as creatures which poffeffed a real exiftence. Readers, for their part,
were glad to believe all that was ftriking and awe-
infpiring. They, no more than authors, dreamt
of weighing authorities.

Turning to Greek writers or retailers of natural
hiftory, Homer and Hefiod alluded to many fables,
and mentioned many plants and animals in words
which fucceeding Greek writers feized upon and
amplified. Hippocrates, B.C. 460, may be termed
the firft regular writer of natural hiftory, although
much has been attributed to him which belongs to
writers of the fame name. Ariftotle, B.C. 356, is
fuperior to all other Greek writers in copioufnefs
and method. Several of his treatifes on natural
hiftory have been loft, but what remains gives a
high idea of his fagacity. His royal pupil Alexander is faid to have fent him fpecimens from the
Eaft. Theophraftus, B.C. 322, has left behind
valuable writings on botany. Strabo, B.C. 30, is
ufeful for geography. Ctefias, who was a contemporary of Herodotus, wrote on the producls
of Perfia and India. Xenophon's work on the
chafe was fupplemented by Arrian's book at the
beginning of the fecond century after Chrift.

by Diofcorides. Paufanias, A.D. 160, touches on
much that is of phyfical and economical intereft
in his "Itinerary of Greece." The "Onomafticon"
of Pollux, a Greek fophift and grammarian, A.D.
183, treats in ten books of the meals, hunting,
animals, etc., of the ancients. Oppian and Ælian,
in the beginning of the third Chriftian century, are
of confiderable intereft to the ftudent of natural
hiftory. The former is the author of a long poem
on "Fifh and Fifhing," and another on "Hunting
and Dogs," both of which difplay the characteriftic
want of accuracy of the ancient zoological writers.
Among other works, Ælian wrote feventeen books
"De Animalium Natura." Thefe have come down
to us. They are feemingly thrown together without any definite arrangement, and abound in
hearfay and marvellous anecdotes. Of Stobæus,
beyond the fact that he was born at Stobi, in
Macedonia, little is known. Even the time at
which he lived is uncertain. He and Photius,
however, have refcued for us numerous interefting
details of Greek life and many extracts from earlier
writers. Among thefe authors, then, the ftudent
of Greek natural hiftory has to quarry.

In the century before our Lord, Cæfar and
Varro among Latin authors claim attention. The
former contains much that is valuable, efpecially
in relation to Gaul and Britain; the latter, of
large and varied erudition, wrote no fewer than
490 books. His three books "De Re Ruftica"
are the moft important treatifes extant upon
ancient agriculture. Book I. treats of farms and

lands ; Book II. of the management of cattle ;
Book III. of the fmaller animals of a farm—hares,
dormice, etc. The poem (afcribed to Ovid) on fifhing ("Halieuticon"), is merely a fragment, yet contains many fpirited lines, and the wiles of the *lupus*
to efcape from the hook as there defcribed are the
fame which have frequently been experienced by the
modern falmon and trout fifher, when the fifh—

> "in auras
> Emicat, atque dolos faltu deludit inultus."

The vaft compilations of Pliny, A.D. 79, avow-
edly intended for a book of reference, have proved
a mine of wealth to all fucceeding writers on
natural hiftory. They are very uncritical; Pliny's
chief anxiety apparently having been that no moment fhould be wafted, and that everything which
he heard fhould at once be reduced to writing.
Nemefianus wrote on hunting, fifhing, and navi-
gation. Some three hundred lines only of his
poem on the firft of thefe fubjects have been pre-
ferved. Much that is interefting may be found
in Martial's "Epigrams." Mr. Simcox fpeaks of
the "careffing defcriptions" of Apuleius, A.D. 163 ;
a few pearls may be collected from the depths
of his rhetorical fea. Juvenal here and there, in
his gloomy pictures of Roman fociety, throws in
a brighter tint which he has felected from what
may be called the natural hiftory of his day. All
fcholarly fifhermen know that charming idyll of
Aufonius on the Mofelle. He was evidently an
angler, to judge from the fpirited and life-like defcriptions of fifh and fifhing which he introduces.

Befides thefe authors, the ordinary poets have been freely laid under contribution in the following pages. They form a fample of the wealth of material which yet remains for zoologifts in the writers of Greece and Rome.

For much of this brief account of Latin authors I am indebted to Mr. Simcox's "Hiftory of Latin Literature" (Longmans).

M. G. W.

GLEANINGS FROM THE NATURAL HISTORY OF THE ANCIENTS.

CHAPTER I.

A HOMERIC BESTIARY.

IN fpite of the attention which has of late years been devoted to Homer, very little care has been expended on the plants and creatures which he introduces in his two immortal poems, and yet the fubject is replete with intereft. From the manner in which he notices the moft ftriking features of the flora of Greece, or the remarks which he makes on animated nature, fomething of the man's perfonality and taftes might, it is only reafonable to fuppofe, be inferred. The attempt to recover fpecial traits of the poet by this method, however, fails, and we are reduced, did we only judge by this line of argument, to fall back upon the view of thofe critics who hold that the "Iliad" and "Odyffey" were fimply a floating collection of

B

ballads put together by Peififtratus, while no actual Homer ever exifted ; or, at all events, that he never wrote the fragmentary verfes which were thus pieced together. Interefting queftions alfo arife refpecting the conformity of the Homeric fauna and flora with the prefent ftate of Greece ; what animals or birds have become extinct or diminifhed in numbers ; whether any remains of the prehiftoric condition of the country are apparent in the poems and the like. Unluckily the evidence for thefe facts within the Homeric poems is very fragmentary, and there is an utter want of authorities with which to compare their ftatements until the time of Herodotus is reached. A fplendid proceffion indeed of animals fet in a beautiful landfcape is prefented to our eyes in Homer, much as the vifitor to an Egyptian temple gazes at the painted birds, beafts, and trees on its walls. But the mind muft for the moft part deal with thefe reprefentations as if ifolated from all further knowledge of them. In too many cafes, too, Homer only introduces his birds and animals by way of fimile. They are not defcribed as a natural hiftorian would depict them ; they are hinted at and alluded to. So that the ftudent of Homer's natural hiftory finds himfelf baffled on every fide.

Yet a few curious facts emerge on careful inveftigation. The predominance of the lion with Homer in fimiles ferves to fhow that this animal was familiarly known in Europe in his time. For many centuries there have been no lions in this continent. The three chief varieties of the animal

at prefent are the Barbary, Senegal, and Perfian
lion.   The difappearance of the lion before
civilized life and agriculture is only fecond to that
of the elephant.   Lions have died out in Egypt,
Syria, and Paleftine as well as in our continent,
and are being driven farther and farther into the
tracklefs wilds of South Africa as population
fpreads up the river valleys, and graffy flopes are
enclofed for farms.   Herodotus tells us that lions
abounded on the rocky portions of Macedonia
and Theffaly.   They attacked the baggage animals
of Xerxes on his march through thefe diftricts into
Greece, and fell fpecially upon the camels.   The
hiftorian naïvely wonders at them for abandoning
their ordinary habits of preying on horfes, oxen,
and men to attack camels, a creature which they
could never before have feen.   He gives a moft
valuable notice, too, of the region haunted by
thefe lions, which, it feems, was from the river
Achelous (the prefent Afpro Potamo) in the weft
to the Neftus or Mefto in the eaft, the boundary
between Thrace and Macedonia.[1]   As fhowing
the tendency of the ancient natural hiftorians to
copy one another, it is worth remarking that
Ariftotle and Pliny, when treating of lions, give
the fame limits for them.   Cybele's chariot was
reprefented as drawn by lions; another teftimony
that the early Greeks knew the character of the
localities frequented by thefe animals.   Of Ariftotle's two kind of lions, the thicker and more
hairy variety feems to refer to the ordinary

[1] Bk. vii. 125, 126.

African lion with fine flowing mane; the other, which he defcribes as longer in fhape and more ftraight-haired, might mean what is now known as the manelefs lion of Gujerat.[1] An amufing chapter of Aulus Gellius[2] arraigns Herodotus, " the moft noble of hiftorians," for ftating that the lionefs only brings forth once during her life, and then only one cub, giving the marvellous reafon which may be found in the third Book of Herodotus, the laceration of the mother's internal membranes by the fharp claws of the cub. Againft this teftimony he quotes paffages of Homer, " the moft illuftrious of poets," to fhow that lions defended their cubs, not their cub ; and continues by quoting Ariftotle on the point, who calls it " an old woman's fable." But we incidentally learn that lions had become fcarcer in Ariftotle's time, a hundred years after Herodotus, as the former fays, " The ftory hath been put together from the fact of lions being fcarce, and the inventor of the myth not knowing how to account otherwife for this fact." Another inftance of credulity immediately fucceeds this difcriminating remark, which alfo fhows the utterly uncritical ftate of mind of the ancients, even of fo diftinguifhed a philofopher as Ariftotle, when the weighing of evidence and collection of facts, which is fo rigoroufly exacted by the modern inductive philofophy, is concerned. " The Syrian lions," he fays, " bear at firft five cubs, next year four, and fo on down to one, after which they never again generate."

[1] Ar. ; "De Anim. Hift.," ix., 31.   [2] *Ibid.*, xiii. 7.

Agamemnon wears a lion's ſkin as a mantle;
but the animal generally appears in ſimiles.
Penelope ponders on her bed before ſleeping, as a
lion when ſurrounded by a ring of hunters takes
counſel with himſelf. We ſee the lion in ſuch
paſſages exulting at finding prey, whether ſtag or
wild goat, killing a hind's fawns, putting to flight
and ſeizing oxen, terrifying bleating goats by
his preſence, driven ravening by men and boys
from the fold, ſlaying a bull, fighting with a wild
boar for water with its cubs, or tracking out
a man who has ſtolen them, being attacked and
killed by angry villagers, or itſelf attacking the
folds. Each of theſe pictures is beautiful in itſelf,
and the whole give an excellent hiſtory of the
habits of the European lion. Odyſſeus, after the
ſlaughter of the ſuitors, glares round him like a
lion. Lions were engraved on the belt of Hercules,
and ſurrounded the ſorcereſs Circe's abode; cats
even at this early period being favourite animals
of witchcraft. Proteus again changes himſelf into
a lion, ſo that this animal muſt have been ſufficiently familiar to Greeks. When the ſavagery of
Cyclops devouring the two hapleſs comrades of
Odyſſeus has to be painted, Homer makes him "eat
like a lion from the mountains," tearing them
limb from limb and not even leaving their bones.
Jackals are only introduced at any length in one
paſſage, but that an eminently characteriſtic one.
The Trojans follow Odyſſeus " like dappled jackals
from the mountains ſtanding round a wounded,
branchy ſtag, whom a hunter has ſmitten with an

arrow. It efcapes by fpeed of foot while its blood is warm and its knees are firm, but when the bitter fhaft fubdues it, then ravening jackals tear it to pieces in a fhady grove among the hills ; but the deity brings there a mighty lion, when they fhrink afide while he devours."[1] The panther only of the *felidæ* is mentioned befides the lion. Paris and others wear its fkin. Its fiercenefs is prominent in a fimile when it is reprefented as iffuing from thick covert to charge the hunter, in no way difmayed at his prefence or at the baying of the dogs, and attempting to ftrike him down. Even when pierced by his fpear it ceafes not its rage until overwhelmed by darts or done to death. The lefs warlike tone of the "Odyffey" is indicated by the fact that there are only four fimiles in it taken from the lion, whereas there are eleven in the "Iliad." The vulture only appears once, war never, and ftorm never.[2]

On the mighty belt of Hercules, in Hades, were wrought bears, the only evidence that Homer knew that animal. This is the *urfus Arctos*, once an inhabitant of our own iflands, and ftill to be found in certain mountainous diftricts of Europe. The wild boar is much more familiar to Homer ; it was facrificed to Zeus and the Sungod, and alfo appears in the belt of Hercules. Proteus tranfforms himfelf into it. The Calydonian wild boar roots up trees in a mythical fafhion, fuggeftive of fome dim remembrance of the mammoth.[3] The

[1] "Iliad," xi. 474.   [2] Gladstone ; "Juv. Mundi," p. 514.
[3] "Iliad," ix. 535.

dogſkin helmet of Odyſſeus is adorned with teeth of a wild boar. Two warriors fall upon the foe like two wild boars ſtoutly charging the hounds. The following pictures are ſo lifelike that it is hard to conceive that Homer had not witneſſed them. " As when a boar upon the mountains, truſting in his ſtrength, abides the mighty on-coming ruſh of men in a lonely place, and the briſtles riſe erect upon his back while his eyes ſhine with flame ; but he gnaſhes his teeth, eagerly deſirous to avenge himſelf on dogs and men, ſo did Idomeneus," etc. And again : " They ruſhed forwards like hounds which ſpring upon a boar, after he has been wounded, in front of youthful hunters."[1] Another vivid picture repreſents a lion and a boar fighting for a rill of water on the mountain-tops, and the lion ſubduing the panting boar.

The word " elephant " is only uſed by Homer for a diſtinctively eaſtern product, ivory. Bulls were found in a wild ſtate on the Greek mountains, as until recent centuries in our own land. Their hides were uſed for ſleeping on. An alluſion occurs to an active hunter cutting down a wild bull by a ſtroke behind the head with a ſharp axe. Scamander is ſaid to roar like a bull. When Penelope unlocks the doors of her treaſury, as they roll back they roar like a bull feeding in a meadow. Oxen were, of courſe, domeſticated from very early days. Laomedon cauſed Apollo to feed " his heavy-footed, crumpled-horned oxen in the lawns of many-valed wooded Ida."[2] Oxen

---

[1] " Iliad," xiii. 471 ; xvii. 725.   [2] *Ibid.*, xxi. 448.

were eaten at a funeral feaft, and facrificed, efpecially black ones, to Pofeidon.   The mifguided
followers of Odyffeus perifhed through their folly
in eating the oxen of the Sungod, in the Ifle of
Thrinacia, when the Sungod amufingly threatens
Zeus that if the facrilege be not avenged, he will
go down to Hades and fhine among the dead.[1]

Wild goats feem to have been found on lonely
mountains.   In the ifle off the land of the Cyclopes
were herds of them.   They were eaten at feafts.
We find them fhrinking with fear from a lion.
Argus had been ufed to hunt them.   The horns of
one are mentioned as being fixteen palms in length,
which were made into a bow and tipped with gold.
Two fpecies of wild goat yet inhabit Europe, the
Capra ibex of the Alps, whofe horns will meafure
two feet eight inches in length; and the *C.
Pyrenaica*, of which the horns are only two inches
lefs.   The goat was facrificed to Apollo in the
Homeric poems.   A fimile in the " Iliad " reprefents two lions as fnatching away a goat from
fharp-toothed dogs; they bear it off in their jaws,
raifing it on high from the earth among the
thickets.

The Homeric dogs much refemble modern dogs
in their habits.   They tear corpfes (like the dogs
of Eaftern cities and countries) in conjunction with
the fowls of the air; and guard fheep and fwine.
Eumæus, the fwine-herd, thus employs four.
They hunt lions, boars, ftags, roedeer and hares.
A characteriftic paffage defcribes their behaviour

[1] " Odyffey," xii.

with lions. It forms a compartment in the fhield which Hephæftus forged for Achilles. " On it he fafhioned a herd of ftraight-horned kine ; the cows were made of gold and tin, and with lowing they ran forth from the ftall to their pafture by a rufhing river edged with rattling reeds. Four fhepherds of gold marched along with the kine, and nine fwift-footed dogs followed them. But two monftrous lions among the leading kine feized the loud-roaring bull, and he, mightily bellowing, was dragged along, while the dogs and youths followed them up. They, however, having torn off the hide of the great bull, proceeded to lap up its bowels and black blood; but the fhepherds fruitleffly preffed upon them, urging on the fwift dogs. They, indeed, kept on fpringing back in difmay from an attempt to bite the lions, but ftanding very near continued howling and avoiding them."[1] No greater reproach can be addreffed to a warrior than to ftigmatize him as poffeffing " a dog's eye and a ftag's heart." Dogs bay round a palace in Ithaca and tear intruders, juft as the Moloffian dogs of old and prefent days refent the approach of ftrangers. Telemachus ftalks about his ifland home like a modern country gentleman, with his dogs following him. The epifode of Argus, the faithful dog of Odyffeus, is too well known to need more allufion to it.[2] In the palace of Alcinous were hounds of gold and filver, the work of Hephæftus; to heighten their marvel the poet, as often in the fhield of Achilles and elfewhere,

---

[1] " Iliad," xviii. 581.   [2] " Odyffey," xvii.

reprefents them as being animated. Here they are "immortal and free from old age for aye."[1] Another celebrated dog of myth was the dog of Hades, afterwards known as Cerberus. When Odyffeus meets the fhade of Heracles in the lower world, the latter tells him that he had been compelled to enter Hades while he was yet alive, and drag this dog to the upper air, " for no greater tafk could be devifed ;" but Hermes and blue-eyed Athene helped him to perform it.[2] Orion's dog was a well-known ftar. In all thefe cafes the dog is even in Homer's time a familiar domeftic creature. Lap-dogs too are named. Perhaps a faint reflection of the wonder which the taming of the creature firft caufed among men yet glimmers on the mythical ftories juft related.

Stags and fawns are frequently mentioned in the Homeric poems ; this is only natural, confidering the numbers which in the early days of Greece muft have been found on her mountains or feeding in the fair glens befide them. Sheep appear among domefticated animals bleating as they wait to be milked. The riotous wooers of Penelope eat fheep, kine, and goats. There is yet a wild fheep in Sardinia, known as *ovis mufmon*, with horns one foot eleven inches long. Lambs are born with horns, fays Homer, in Libya, and the fheep there bring forth thrice in a year.[3] Can this ftory of the horned lambs be a reflection of the true hiftory of the wild fheep of Europe? White

---

[1] "Odyffey," vii. 94.   [2] *Ibid.*, xi. 263.
[3] *Ibid.*, iv. 85, 86.

and black lambs were facrificed to Earth, Apollo, Helios, and Zeus. As for fwine, the herds kept by Eumæus, their huge pigfties, their grunting, and the manner in which one is butchered by Odyffeus, are amufingly related in the fourteenth Book of the "Odyffey." They are called "delicately fed," and they were finged when killed for a feaft.

The wolf was well known to the early Greeks. We find it in conjunction with lions roaming round the mythical palace of Circe. It rufhes on lambs and kids, like champions hurrying to the din of battle, and preys in conjunction with pards and jackals upon ftags. The myrmidons whom Achilles leads to war are compared to a flock of wolves in a fine naturaliftic picture; "like wolves, ravening after prey, around whofe hearts is unfpeakable ftrength, which, having pulled down a mighty horned ftag in the mountains, tear it to pieces; and the face of them all is red with blood. Then they rufh off in a flock to lap up the furface of the dark waters from a black-flowing fountain with flender tongues, vomiting forth clotted gore, and their courage within their breafts is dauntlefs, and their ftomach is diftended."[1] We hear of a wolf-fkin as well as a dog-fkin helmet, and of one made of a weafel's or more probably a marten's fkin.[2]

The horfe is conftantly mentioned, but never feemingly as an animal to be ridden. A characteriftic paffage, the only one in which the animal

[1] "Iliad," xvi. 156. [2] *Ibid.*, x. 335.

is named, introduces the afs: "As when a fluggifh afs, paffing by a cornfield, hath overborne the boys, and many a cudgel has been broken round his fides, but he, entering in, ravages the deep crop while the boys beat him with fticks. Yet their ftrength is but feeble, and hardly have they driven him out when he hath taken his fill of the grain."[1] Mules were apparently much efteemed. There is a mention of them as being very ftrong and employed in dragging heavy beams; they draw Priam's chariot, having been given him as illuftrious gifts by the Myfians. When Naufícaa takes her garments to be wafhed by the fea-fhore, they are drawn thither in a waggon by mules.

The lift of mammals in the two great Homeric poems, is completed by the hare, which is reprefented as torn by an eagle, as in the fplendid chorus at the beginning of the Agamemnon of Æfchylus, and by feals. A very curious paffage relates how Menelaus, thanks to the help of Eidothëe, daughter of Proteus, furprifed that "old man of the fea" among his feals which flept around, "exhaling a bitter fmell of the deeps of the fea." The ftench of thefe animals is again defcribed as being overpowering, until the goddefs luckily bethought herfelf of rubbing a little ambrofia under the nofe of each man, which effectually removed the ill favour.[3] The poet probably alluded to the *phoca monachus* of the Mediterranean, or perhaps the *phoca vitulina* alfo feen at times in

---

[1] "Iliad," xi. 557.   [2] *Ibid.*, xxiv. 277.
[3] "Odyffey," iv. 404, 436.

that fea. Seals are very numerous in the Cafpian Sea, and are even found in the falt fea of Aral, as well as in the frefh-water loch of Baikal.

In a very curious paffage of the "Odyffey" (xxiv., 6) reprehended by Plato in his "Republic," the fouls in Hades are compared to bats, "which fly fqueaking in the recefs of a marvellous cavern, when one has fallen from the rock out of the clufter," and Odyffeus clings to the fig-tree of Charybdis like a bat.

Thus a Homeric houfehold kept the fame domefticated animals as we do at prefent—horfe, afs, mule, fheep, oxen, pigs, and goats. It is fingular, when the origin of the domeftic fowl is remembered (the jungle fowl of India), that fowls are not named in the Homeric poems. The common fuppofition that thefe birds were brought weftward by the primitive Aryans, feems therefore erroneous. They came through hiftoric intercourfe with the Eaft, and in Homer's time there is plenty of evidence to fhow that this intercommunication of Europe and Hindoftan had not yet begun.

Turning to reptiles, Proteus turns himfelf into them,

$$\text{ὅσσ' ἐπὶ γαῖαν}$$
$$\text{ἑρπετὰ γίγνονται.—}Odyffey, \text{ iv. } 417.$$

A dragon is one of thefe fhapes. The dragon (or ferpent) is reprefented as eating birds in other paffages; caufing a man to fhrink back as he meets it in his path; an augury appears of "a high-flying eagle, on the left hand, dividing the people, bearing a monftrous bleeding ferpent in its

claws, alive, yet gafping; and not yet had it forgotten to fight, for it fmote the eagle which held
it in the breaft by the neck, bending itfelf back
to do fo. But the other let it drop to the ground,
grieved at the anguifh, and caft it down into the
midft of the crowd, while it fled fcreaming on the
wings of the wind." Snakes (or dragons) of
"Cyanus" are fafhioned gleaming like rainbows
on Agamemnon's fhield. A dying man lies like
a worm;[1] while maggots, in another paffage, are
made to eat corpfes.

With regard to fifh and fifhing, fome fingular
facts appear in the Homeric poems. We will
group them together without entering into modern
views of claffification, feeling fure that Homer regarded the whale, for inftance, as a fifh, and not
a mammal. Fifhers apparently cruifed from ifland
to ifland of the Ægean, for bodies of the flain
wooers are delivered to the fifher-folk to be conveyed each to his own city in fhips. The whales
(or larger fifh of the fea) are faid to fport round
Pofeidon's chariot as he drives over the fea to
recognife their king. A fea-monfter (or κῆτος,
which means any large fifh or monfter) purfued
Hercules from the fhore of the Troad to the plain
in the myth. Fifhes, and efpecially eels,[3] are
feveral times fpoken of as devouring the flain.
Dolphins purfue and eat fifh. Homer had noticed
a fifh "rife," though it is fomewhat bewildering

---

[1] "Iliad," xiii. 654, σκώληξ ; *ibid.*, xxiv. 414, ὐυλὰι.
[2] "Odyffey," xxiv. 418.
[3] "Iliad," xxi. 353 (the eels and fifh in the river Xanthus).

to find out what the following paſſage means. A hero is ſtricken and falls inſenſible, "as, under the ripple cauſed by the north wind, a fiſh leaps up on the weedy ſhore, and the dark wave covers it."[1] Probably it would be better rendered "a fiſh leaps up by the weedy ſubmerged reef." As for the capture of fiſh, the Læſtrygons hurl rocks at and kill the haplefs mariners of Odyſſeus, and, "like men ſpearing fiſhes," they bear home their frightful meal. But angling was known to Homer; "as when a man ſitting upon a projecting rock draws a ſacred" (*i.e.* mighty) "fiſh to land from the ſea with line and ſhining brafs" hook.[2] The fiſhing-rod is not here named, and the "brafs" hook was probably a hook of bronze, one of which is figured in Evans's "Bronze Implements." But in the "Odyſſey" (and this ſeems a confirmation of the view that it is a later poem than the "Iliad") a rod is employed; "as when a fiſherman on a projecting rock, with a very long fiſhing-rod letting down his baits as a ſnare to the little fiſh, flings into the ſea a horn of an ox of the homeſtead, and then, as he has caught the fiſh, flings it gaſping on the ſhore."[3] Here a difficulty is contained in the uſe of the horn. It was probably a ſheath coming over the bait, either to prevent its being waſhed off, or to protect it from crabs and the like.

Theſe are the two chief authorities for fiſhing

---

[1] "Iliad," xxi. 692. [2] *Ibid.*, xvi. 406.

[3] "Odyſſey," xii. 251. A cognate paſſage occurs in the "Iliad," xxiv. 80 : Iris "plunged into the depths of the ſea like a leaden plumb which in the horn of an ox of the ſtall entering the ſea drops through it, bearing death to ravening fiſhes."

in the Homeric poems.   The monfter Scylla is
faid to fifh, with her hands groping to catch dog-
fifh or dolphins.   Plato notices that the Homeric
heroes in their feafts never eat fifh, and that their
viands are always roafted, never boiled.   It is a
curious confirmation of the former ftatement that
when   the   men   of   Odyffeus   fifh   in   the   Ifle
Thrinacia, with "crooked hooks," for fifh or fowl,
under the preffure of famine, their mafter will
have   nothing   to   do   with   it,   but   wanders   off
alone.[1]   Yet in a picture drawn by the hero of
a righteous and profperous king, one touch is that
"the fea for him gives fifh."[2]

A fingular paffage occurs in the "Odyffey,"v. 432,
where Odyffeus is compared, while in danger of
drowning, to a cuttle-fifh "which is dragged out
of its hole, the many pebbles clinging to its
fuckers ;" juft in the fame manner the hero's fkin
is torn off from his hands as he grafps at the
rocks, and the mighty wave covers him.   Again,
a man ftricken with a mortal wound, who falls
headlong from his chariot, is jeered at in the
" Iliad "—" if only he were in the fifhy deep, this
man   would   fatiffy   many   men   by   grafping   for
oyfters, plunging in from a fhip, although it was
ftormy weather."[3]   Were it not for thefe curious

[1] Plato, "Repub.," 404, B. ; "Odyffey," xii. 331.
[2] " Odyffey," xix. 113.   That fifh were eaten, too, appears
from Od. xxii. 383, where Odyffeus fees the flain wooers lie
" like fifh which fifhermen have drawn from the grey fea in a
many-mefhed net to a hollow beach, and they all longing for
the fea-waves are heaped upon the fand, and the fun fhining on
them takes away their life."
[3] "Iliad," xvi. 745.

words, we ſhould not know that oyſters were a dainty ſo early as the Siege of Troy.

The zoology of the Homeric poems may be completed by a glance at the inſects, etc., which are named by the poet. The "glancing gadfly" attacks the herds. One kind of worm or weevil attacks the wood of Odyſſeus's bow; another eats corpſes. Locuſts are repreſented fleeing from fire. Flies are often mentioned. A little one perſiſtently attacks a big man, in one paſſage; in another, flies hum round the milk-pails in ſummer, or round the ſhepherd's pen. A beautiful ſimile repreſents Athene cauſing an arrow to fly off from Menelaus " as a mother drives off a fly from her child when enjoying a ſweet ſleep." Still more celebrated is the paſſage which introduces the favourite Greek inſect, the chirping *tettix;* the old men of Troy are no longer able to fight, but are "excellent talkers, like tettixes" (graſſhoppers), " which, in the thickets, ſitting on a tree, ſend forth a thin clear voice."[1] Spiders even had been noticed by Homer, and were not deemed by him, any more than Shakeſpeare deems the toad, unworthy the dignity of poetry. The fetters which Hephæſtus conſtructed in order to enſnare his erring wife were fine, yet ſtrong as ſpider's web.[2] Round the neglected bed of Odyſſeus were foul ſpider-webs. Bees are mentioned as neſting in a hollow rock, not a beehive—another evidence of the antiquity of theſe poems. Evidently bees had not yet been domeſticated. They made their

[1] "Iliad," iii. 151.   [2] "Odyſſey," viii. 280 ; xvi. 35.

nefts, too, in the hollow cave at the landing-
place in Ithaca. Wafps are named with them as
making their abodes in a rugged path, and not
quitting them at the approach of the fpoiler, but
fighting for their young. A paffage which fpeaks
of the Trojans iffuing forth from their city, fhows
that boy-nature was with the Greeks much the
fame as it is with us; "they poured out like wafps
dwelling near a road-fide, which filly boys are accuftomed to irritate, ever difturbing them as they
live in their road-fide homes, and caufe a common
evil to many; and if by chance a wayfarer going
by fhould unwittingly difturb them, they with their
ftrong hearts fly each one ftraight before it, and
fight for their little ones."[1]

In contemplating the wide range of Homer's
natural hiftory, and the evident love with which
he dwells upon fome of the nobler forms of
animal life, we cannot help being ftruck with the
prodigality of his allufions to animals. It fhows
the ftrength of his fympathies with outer nature.
He may thus be advantageoufly compared with
his fucceffors in Epic poetry. Virgil lavifhes his
tendernefs on birds and beasts in the "Eclogues"
and "Georgics," but feldom names them in the
"Æneid;" feldom, that is, as compared with the
frequency with which they do duty as fimiles, or to
enliven the Homeric landfcapes. Save in his firft
book, or when treating more efpecially of creation,
Milton is equally reticent. Indeed, the few allufions which our poet does make to animal life,
or even to plants and flowers other than thofe

---

[1] "Iliad," xvi. 259.

ſuggeſted by his claſſical models, are ſomewhat ſurpriſing in one ſo fond of Engliſh landſcape, as we
know him to have been from his "Penſeroſo" and
"Allegro," and from the records of his home-
life which have been preſerved. His ſuſceptibility
to muſic was extreme, and his gorgeous deſcriptions of muſical harmony, hymns, and the like,
have often been noticed. But it has not been
remarked hitherto that his ear, rather than his eye,
caught thoſe reflections of nature which he has
loved to reproduce in deathleſs verſe. The crow-
ing of the cock, ſinging of the lark, warbling
of the nightingale, and ſimilar ſounds at once
occur to the memory. This may account for the
paucity of his notices of animated nature. The
cuſtom of welcoming the ſounds, and ſongs, and
cries of external nature through the ear muſt often
have mercifully ſtood him in good ſtead when the
affliction of blindneſs fell upon him in late life.
It is obvious how diſtinct from both Virgil and
Milton is Shakeſpeare in the manner he enlarges
upon and welcomes into his verſes the flowers,
birds, and beaſts of common life. Here, as alſo
in his graſp of human greatneſs, and his delinea-
tion of the maſter-ſprings of action, he can only
be compared with Homer. Both together are the
moſt catholic of poets, in the depth of their
ſenſibilities, the range of their inſight, and the
power and far-reaching graſp of their ſympathy.
The natural hiſtory of Shakeſpeare has been and
ſtill is ſtudied from every point of view ; the above
is at leaſt a humble contribution towards the fuller
enjoyment of Homer.

CHAPTER II.

GREEK AND ROMAN DOGS.

"Certes, the longer we live, the more things we obferve and marke ftill in thefe dogges."—PLINY, *Nat. Hift.*, viii. 40 (Holland).

THE Greeks and Romans were acquainted with the virtues of the dog, and valued it for its ufe in hunting and the care it took of the flocks or of the houfe, but ufually regarded it, much as did the ancient Hebrews, as a type of fhamelefs and audacious evil. So Helen, in the depths of her felf-abafe-ment, applies the comparifon to her own life in the "Iliad," and Hecuba, according to the myth, was changed into a dog. Wealthy men and kings had lapdogs, indeed, but took none of that pleafure in the affection and faithfulnefs of a fagacious animal which caufes the dog to be fo highly prized in modern life.[1] In augury dogs were unlucky, bafe animals (*obscœnæ canes*—"Georg.," i. 470), and

[1] See a noble paffage on the difference between claffical and Chriftian appreciation of Nature in Rufkin's "Modern Painters," vol. ii., p. 17.

Horace naturally introduces the dogs of the Suburra, the "artificers'" quarter and the moſt abandoned precinct of Rome, in a witchcraft ſcene of cruelty and uncleanneſs (Ep. v. 58). The moſt important ſtar in the conſtellation of the dog was Sirius; "about four hundred years before our era, the heliacal riſing of Sirius at Athens, cor-reſponding with the entrance of the ſun into the ſign Leo, marked the hotteſt period of the year, and this obſervation being taken on truſt by the Romans of a later epoch without conſidering whether it ſuited their age and country, the *dies caniculares* became proverbial among them, as the dog-days are among ourſelves, and the poets con-ſtantly refer to the lion and the dog in connection with the heats of midſummer."[1] By way of con-tempt, the worſt throw at the dice was known among the Latins as canicula, juſt as we brand bad Latinity as dog Latin. The porter at the entrance of both Greek and Roman houſes was uſually attended by a dog; hence the expreſſion *cave canem*, which was proverbial among the Romans. Sometimes a painted dog with the warning was employed, as in a houſe which has been opened at Pompeii.

Greece and Rome do not appear to have known as a diſtinct breed that peculiar lightly built type of the family, like a greyhound, which was com-mon in Egypt. It had much affinity both in cha-racter and derivation to the jackal. Dogs are not unfrequently found repreſented on the Babylonian

[1] "Dictionary of Antiquities," Art. "Aſtronomia."

cylinders, and one kind of dog is of this fame greyhound type, while the other, known as the Indian dog, refembled our maftiff.[1] The excellence of the Spartan hound is often celebrated by the ancients, while the Moloffi in Epirus poffeffed a breed of large dogs which was, if poffible, ftill more renowned. Mr. Hughes, in his travels through Albania, found thefe dogs as numerous and fierce as they were in old days. The breed, he thought, had in no refpect degenerated. He defcribes them as " varying in colour, through different fhades from a dark brown to a bright dun, their long fur being very foft, and thick, and gloffy. In fize they are about equal to an Englifh maftiff; they have a long nofe, delicate ears finely pointed, magnificent tail, legs of a moderate length, with a body nicely rounded and compact."[2] Ariftotle, fpeaking of thefe dogs, fays that a difference of qualities is obfervable in the males and females, the latter being more gentle and tractable, and more eafily taught. Therefore the females are more prized among the Spartan hounds as being of a nobler nature than males. The Moloffians, he obferves, are not better hunting dogs than others, but form excellent fheep-dogs, from their fize and courage in attacking wild beafts.[3] In another place he gives an excellent life-hiftory of dogs, their generation, birth, dentition, and the like; " moft dogs," he adds, " live

---

[1] Rawlinfon's "Ancient Empires," ii., p. 494.
[2] Arnold's "Rome," ii., p. 438.
[3] "Hift. An.," ix. 1.

about fourteen or fifteen years, but fome twenty; wherefore fome think that Homer was quite correct in making the dog of Ulyffes die in his twentieth year."[1]   He had noticed, too, that dogs dream, from their howling in fleep, as if they were then following the chafe.   We believe it, however, to be a kind of nightmare when dogs thus moan in fleep, in fpite of the Laureate's words—

"Like a dog he hunts in dreams."

Pliny's account of the dog may be here fummarized.[2]   Along with the horfe he is the moft faithful of animals to man.   A dog has been known to defend his mafter from robbers as well as he was able, and on his protector being flain, to have watched his body, driving birds and wild beafts from it.   Another dog in Epirus, on meeting his mafter's murderer, by barking and biting compelled him to confefs the crime.   Two hundred dogs accompanied the king of the Garamantes from exile, ranging themfelves in warlike order againft all adverfaries.   Some nations have had armies of dogs, which never declined a combat, and never clamoured for pay.   When the Cimbri were flain, their dogs defended the waggons of the tribe. When Jafon the Lycian was killed, his dog refufed to take food, and died of grief.   Dogs have been known to throw themfelves into the flames when the funeral pyre of their mafters was kindled. He gives feveral other inftances of the dog's faithfulnefs and gentle domeftic habits.   A dog will

[1] "Odyffey," vi. 20.   [2] "Hift. Nat.," viii. 40.

remember long journeys, and his memory is more
retentive than that of any other creature fave man.
A dog's attack and rage may be mitigated by the
perfon fo affaulted fitting down quietly on the
ground.   This belief, as we have fhown, is as old
as Homer.   The Indians are reported to crofs
their dogs with tigers; the firft and fecond families
which refult are condemned as too favage, but the
third generation is trained.   So cunning are dogs,
that in Egypt they run along, lapping the Nile as
they go, left by halting crocodiles fhould find an
opportunity of dragging them in.   When Alexander the Great was on his march to India, the
King of Albania gave him a dog of wonderful
fize.   Alexander, delighted at its appearance, commanded bears, boars, and ftags to be flipped to it;
but the creature lay motionlefs in fupreme con-
tempt, and at the flothfulnefs of fo huge a form
the king's noble fpirit was aroufed, and he bade
the dog be killed.   His friend now fent another
dog of the fame kind to him, with a meffage that
it was only to be matched with lions or elephants,
and not with fmall game.   The dog foon killed
a lion in the prefence of Alexander, and was next
matched againft an elephant.   Firft of all, with
every briftle on its form erected, the dog bayed
and attacked its enemy, firft on one fide, then on
the other, flipping in and avoiding the elephant's
ftroke wherever an opening prefented itfelf, like a
good boxer, until the elephant grew dizzy by per-
petually turning round to defend itfelf, and finally
falling down, fuccumbed to its petty adverfary.

Dogs frequently go mad during the thirty dog-days, and the difeafe muft be counteracted by fowls' dung being mixed with their food, adds the grave hiftorian, or if they be already fuffering they muft be treated with hellebore. According to Columella, if the tip of a dog's tail be cut off within forty days from its birth, it will never go mad. A dog has been known to fpeak by way of portent, juft as a ferpent ere now barked when Tarquinius Superbus was driven from the throne. "The beft of the whole litter is that whelpe that is laft ere it begin to fee, or elfe that which the mother carries firft into her kennel."[1]

Such were fome current Roman beliefs about the dog. No more celebrated dog than Cerberus appears in claffical mythology. Virgil fpeaks of his "three gaping mouths," and calls him "the gate-keeper of hell reclining in his blood-ftained cave over half-eaten bones." Still more particular is the portrait which the wretched Culex, when untimely flain and fent down to Orcus, draws of him—"Cerberus barks at me with loud bayings, on both fides of whofe neck twifted fnakes briftle, and his bloodfhot eyeballs flafh forth a blaze of flame;" and he adds, "Truftful indeed was he who believed that Cerberus was ever mild-tempered."[2] Homer did not know his name, Cerberus, but fpeaks of Hercules dragging into daylight "the dog of mournful Hades," and in the Odyffey Hercules in the Shades himfelf tells the ftory to

[1] Pliny, "Nat. Hift.," viii. 40 (Holland).
[2] "Georg.," iv. 483 ; "Æneid," viii. 296 ; "Culex," 219, 269.

Odyſſeus—" Zeus enjoined on me hard adventures, yea, and on a time he ſent me hither to bring back the hound of hell; for he deviſed no harder taſk for me than this.   I lifted the hound, and brought him forth from out of the houſe of Hades; and Hermes ſped me on my way to the grey-eyed Athene."[1]   The popular view is well expreſſed by Sophocles ("Œd. Col.," 1568), who ſpeaks of "the unconquerable brute who, as the tale runs, ſleeps in the gates of Hades, poliſhed by the entrance of ſo many ſouls, and, untamable guardian that he is, whines out of the grottoes."   The conception of a dog which guarded Hades came to the claſſical nations, together with the fable of Charon and his boat, from the Egyptians.   Orpheus is ſuppoſed to have introduced theſe myths into Greek fancy.   Heſiod is the firſt Greek to mention the name and genealogy of Cerberus, and with him the dog is " unapproachable, open to no ſoothing, ravenous, the brazen-voiced hound of Hades, ſhameleſs and mighty with fifty heads."[2]   After-poets ſpoke of him as three-headed, with ſerpents for his tail and mane.   At length he becomes hundred-headed, and rivals Oriental monſters in prodigality of horrors. Hercules conquered another dog as well as Cerberus, born (like him) of Typhaon and Echidna, the dog of Geryones. It, too, from reſembling the guard of Hades, is ſometimes called Cerberus.

[1] "Iliad," viii. 367 ; and "Odyſſey," xi. 623 (Butcher and Lang's Tranſlation).
[2] Hesiod, "Theog," v. 388.

Now it is remarkable that there are two dogs of hell in the Vedic mythology, as yet unnamed. They guarded the road to Yama, the king of the departed. This fecond Greek dog, generally known as Orthros, is the exact copy of the Vedic Vritha, and Vritha (like Orthros) is connected with the dawn.[1]

It is characteriftic of the mild-tempered Telemachus,

> "Centred in the fphere
> Of common duties, decent not to fail
> In offices of tendernefs, and pay
> Meet adoration to the houfehold gods,"

that Homer reprefents him, and him alone, in the "Odyffey" as being followed wherever he walks by his dogs.[2] Of Odyffeus himfelf the poet ufes a ftriking ufage; "His heart within him barked" as he glared at the proud mifdoings of the fuitors; "as a bitch walking round her tender pups barks if fhe knows not the man who approaches, and is minded to fight, fo did he growl inwardly when he beheld their evil works."[3] Befides Argus, moft claffical readers will remember the dog which barks at the end of Virgil's incantation fcene, and fhows that the fpells have worked upon the forgetful lover, "Hylax in limine latrat." A poem by Gratius Falifcus in the Auguftan age enumerates fome twenty different forts of dogs, but the Britifh, Spartan, and Moloffian dogs were the types beft known to the ancients. Dogs were

---

[1] See Max Müller's "Selected Effays" (Longmans, 1881), vol. i., p. 497.
[2] "Odyffey," xvi. 61, and xx. 145.   [3] "Odyffey," xx. 13.

kept on the Capitoline as guards for the Temple
of Jupiter, and it was told that while thefe raged
at everyone elfe who approached, they fuffered
Scipio Africanus to draw near unharmed night
after night when he was wont to enter the receffes
of the Temple, and confult there with Jupiter
on the deftinies of the State. Dog-men with dog-
like faces and barkings were fabled by the ancients
to refide in North Africa and alfo on the Indian
mountains, along with other monftrofities, fuch as
one-limbed men, men with their heads below their
fhoulders, and the like.[1] Many of thefe reappear
in the marvellous recitals told by the Mediæval
travellers. The Greek name for a helmet fhows
what was the ultimate ufe of a dog, juft as we
have dogfkin gloves. Virgil does not forget to
recommend the dog to the care of hufbandmen:

> " Nor laft, forget thy faithful dogs ; but feed
> With fatt'ning whey the maftiff's generous breed,
> And Spartan race ; who, for the fold's relief,
> Will profecute with cries the nightly thief ;
> Repulfe the prowling wolf, and hold at bay
> The mountain robbers, rufhing to the prey.
> With cries of hounds thou may'ft purfue the fear
> Of flying hares, and chafe the fallow-deer ;
> Roufe from their defert dens the briftled rage
> Of boars, and beamy stags in toils engage."[2]

As is his wont, Ælian gives many ftories of
dogs and curious fcraps of folk-lore. They have
been known, he fays, actually to fall in love with
men ; their affection is extreme, fo when one
Nicias flipped into a furnace his dogs remained,

---

[1] Aul. Gell., vii. 1, 8, and ix. 4, 9.
[2] Dryden, "Georg.," iii. 404.

howling and dragging out bits of his clothing, by which it was found out how he had perished. Indeed, they insensibly acquire the type and habits of their masters. Thus the Cretans are light, supple, and agile, and so are their dogs. The Molossians are like their owners, most courageous, but when once a Carmanian and his dogs' ire are aroused they are most difficult to be appeased. The Hand of Glory and its use to credulous housebreakers has been described in most books of folk-lore. Ælian gives a somewhat kindred receipt by which a thief may silence the fiercest dog; viz., by holding to it a torch snatched from a man's funeral pyre.[1] It were long, however, to dwell on the superstitions and ancient folk-lore connected with the dog. We fear lest any further attempt to do so might be like inviting readers to a *prandium caninum* (to quote a last allusion belonging to the ancient dog); that is, to a teetotal banquet.[2] There are several chapters on the virtues and vices of dogs in Bochart's " Hierozoicon." Patroclus, in the " Iliad," possesses nine lapdogs (κύνες τραπεζῆες), and Achilles sacrificed two of them on their master's tomb (" Iliad," xxiii. 173). At Rome dogs were annually trussed upon forks, and while thus, as it were, crucified, were hung alive upon elder-trees, to deal exemplary justice upon the race which gave no alarm when the Gauls scaled the capitol. It seems, too, that the Romans, like the Chinese, valued the flesh of

---

[1] " De Nat. An.," i. 6, i. 8 ; vi. 53 ; iii. 2 ; i. 38.
[2] Aul. Gell., xiii., 30, 12.

puppies as an edible in old days.　Hence whelps were ſacrificed as an expiatory offering to the gods, following out the philoſophy of ſacrifice that men ſhould offer to the gods whatever they moſt valued.　"And verily at this day," ſays Pliny, "they make no ſcruple to ſacrifice a yong whelpe before it be full a day old ; yea, and at the ſolemn feſtivall ſuppers ordained for the honour of the gods, they forget not this day to ſerve up at the table certain diſhes of yong whelp's fleſh that ſucke their dams."　At the aditiales, the inaugural feaſts of the magiſtrates, the fleſh of puppies was ordinarily ſerved.　Perhaps the curious in ſuch viands in the Weſtern world might even now have no difficulty in procuring puppies, ready dreſſed for cooking, in the markets of Naples.[1]

[1] Pliny, "Nat. Hiſt.," xxix. 4 (Holland).

CHAPTER III.

THERE are few more vexed queftions in the archæology of natural hiftory than the origin of the dog. The fearcher of bone caverns cannot light upon any definite evidence, inafmuch as the fkulls of dogs, wolves, and their congeners are much the fame. The dog family (*canis*) makes its firft appearance in the lower Pleiftocene era, along with wolves, elephants, and oxen. There is no trace of dogs or other domeftic animals having been known to or ufed by the cave-men ; but in the Neolithic age the dog was occafionally employed for food, probably when old and paft his work, a more humane, if lefs heroic, ending to a life of hunting than was that of the worn-out Argus when he once more faw his mafter ("Odyffey," xvii. 326). In a Neolithic barrow, however, at Eyford, Mr. Greenwell found a dog which had been undoubtedly buried along with a woman whofe fkeleton was ftill, like that of the dog, *in*

situ. Its jaw fhowed it to have been about the fize of an ordinary fhepherd dog. The dog was abundantly reprefented in the Norfolk flint mines known as Grime's Graves.[1]

The dog is met as the trufted friend of man when hiftorical times commence ; thus its commonnefs precludes much exact mention of it. Its exiftence was taken for granted. Theory, therefore, flourifhes abundantly in connection with the early hiftory of the dog, and much *à pofteriori* argument. Such gueffes muft be taken obvioufly at their own value. Thus it does not follow that man in his primitive exiftence as a hunter was aided by the fkill and fpeed of dogs, although Pope may find it convenient to fuggeft the notion to our minds by his well-known lines on the " poor Indian " and his dog. Many favage tribes which live by hunting, at the prefent day, never employ dogs. Nor need it neceffarily be fuppofed that the primitive Aryan fettlers in Europe brought dogs with them. Mr. Darwin has paid great attention to the queftion, and as he inclines to believe that different croffings of fome *canis primitivus*, now loft, with wolves and jackals, may account for the exiftence of the numberlefs modern breeds of the dog, few will venture to contravene his fuppofition.[2] " Many European dogs," he obferves, " much refemble the wolf," and all who have interefted themfelves in this queftion muft

---

[1] Greenwell's " Britifh Barrows," p. 736 ; and fee Dawkins's " Early Man in Britain," pp. 87, 217, 304.
[2] See " Plants and Animals under Domeftication," vol. i., cap. i.

have made the fame remark to themfelves with
reference to fome Englifh fheep-dogs, and ftill
more in the cafe of feveral Continental breeds of
large dogs. Profeffor Owen, however, in his
"Britifh Foffil Mammals," afcribes certain canine
bones difcovered in an Englifh bone-cave to *canis
familiaris,* and thefe are probably the earlieft
authentic remains of the Britifh dog. Befides
the numerous varieties common to England and
Scotland, the latter country poffeffes breeds un-
queftionably peculiar to itfelf, as the deerhound,
Skye and Scotch terriers. Sir Robert Sibbald,[1] when
enumerating the quadrupeds of Scotland in 1684,
names the various kinds of dog as being, "cur,
fhepherd's dog, greyhound, beagle, bloodhound,
moloffus or Englifh maftiff, fetting-dog, water-
fpaniel, terrier, *canis Melitenfis,* a Meffin or lap-
dog." Dr. Caius,[2] writing in 1570, had fcarcely
been fo particular to affign each dog to its own
country, faying amufingly enough, when his words
are contrafted with the fporting of the prefent
day: "I cal them univerfally all by the name of
Englifhe dogge, as well becaufe England only, as
it hath in it Englifh dogs, fo it is not without
Scottifhe, as alfo for that wee are more inclined
and delighted with the noble game of hunting,
for we Englifhmen are adicted and given to that
exercife and painefull paftime of pleafure, as well
for the plenty of fiefhe which our Parkes and
Forefts doe fofter, as alfo for the opertunitie and

[1] "Scotia Illuftrata," Edinburgh, 1684, iii. 5.
[2] "Of Englifhe Dogges," 1576 (reprinted 1880), p. 2.

convenient leifure which wee obtaine, both which the Scottes want."

Narrowing our inveftigations to the dogs of our own land, the next information which we obtain comes from Art. Dogs are frequently found reprefented on the Romano-Keltic pottery of England, efpecially on Durobrivan ware. Thefe dogs commonly fall under one of two types; they are large and fierce, like our prefent bull-dogs and maftiffs; or they refemble a fleet, flender hunting-dog, fuch as our greyhound. By comparifon of the forms ftill remaining at the different mufeums on pieces of pottery, fome particulars might be obtained refpecting the various breeds of the early Britifh dog, if we could be fure that the artift did not ufe conventional or imaginary types of dog-life. At this point, too, the well-known paffages in the claffics which refer to the excellence of Englifh dogs come in. The larger and fiercer kinds were much employed both by the Roman fojourners in Britain and their countrymen at home in chafing the wild boar. Shepherd-dogs, too, may have been needed to tend the "magnus numerus pecorum" of which Cæfar fpeaks in our ifland. The luxury of the Roman capital at York would alfo be almoft certain to demand the fmaller breed for pets. Even in the Homeric times Kings kept them ("Odyffey," xvii. 309). Britifh maftiffs were much celebrated amongft the ancients. Martial fays of another kind (xiv. 200):

"Non fibi fed domino venatur vertagus acer,
Illæsum leporem qui tibi dente feret."

Vertagus is ſaid to be a Keltic word, though it ſomewhat ſuggeſts *verto* as its root, a dog which, like a greyhound and retriever combined, would purſue the windings of the hare's terrified flight, and then return when it had ſnapped up its prey, carrying it to its maſter. The *moloſſus* or maſtiff was a word ſoon uſed in a much wider ſenſe than its primitive meaning, (a dog belonging to the Moloſſi), warranted. Virgil's

> " Veloces Spartæ catulos acremque moloſſum "
> (*Georgics*, iii. 405)

is an inſtance of ſuch uſe, while the other, the Laconian dogs, have not been forgotten by our own Shakeſpeare:

> " My hounds are bred out of the Spartan kind."
> (*Midsummer Night's Dream.*)

And he goes on to ſpeak of their " tuneful cry," reminding us of Walton's enthuſiaſtic words: " What muſic doth a pack of hounds then make to any man, whoſe heart and ears are ſo happy as to be ſet to the tune of ſuch inſtruments!" ("Compleat Angler," i. 1.)

Holinſhed[1] inſerts a curious chapter "of our Engliſh dogs and their qualities" in his "Chronicles." " There is no countrie," he ſays, " that maie compare with ours in number, excellencie, and diverſitie of dogs." Of all who have praiſed theſe creatures, Carden writes moſt marvels of them ; " who is not afraid to compare ſome of them for greatneſſe with oxen, and ſome

[1] "Chronicles" (six vols., 1807), vol. i. 386.

alfo for fmalneffe vnto the little field-moufe."
One of Holinfhed's divifions of Englifh maftiffs
is fufficiently amufing: "Some doo both barke
and bite, but the cruelleft doo either not barke
at all, or bite before the barke, and therefore are
more to be feared than anie of the other." The
whole chapter deferves perufal.

Turning to the numerous varieties of our dogs,
it is worth while quoting fome curious facts here
from Mr. Darwin: "The bulldog is an Englifh
breed, and, as I hear from Mr. G. R. Jeffe, feems
to have originated from the maftiff fince the time
of Shakefpeare; but certainly exifted in 1631, as
fhown by Preftwick Eaton's letters. There can
be no doubt that the fancy bulldogs of the prefent
day, now that they are not ufed for bull-baiting,
have become greatly reduced in fize, without any
exprefs intention on the part of the breeder. Our
pointers are certainly defcended from a Spanifh
breed, as even their prefent names—Don, Ponto,
Carlos, etc.—fhow; it is faid that they were not
known in England before the Revolution in 1688;
but the breed fince its introduction has been much
modified, for Mr. Borrow, who is a fportfman,
and knows Spain intimately well, informs me that
he has not feen in that country any breed 'corre-
fponding in figure with the Englifh pointer; but
there are genuine pointers near Xeres which have
been imported by Englifh gentlemen.' A nearly
parallel cafe is offered by the Newfoundland dog,
which was certainly brought into England from
that country, but which has been fince fo much

modified that, as feveral writers have obferved, it does not now clofely refemble any exifting native dog in Newfoundland." [1]

With regard to this variety of canine breeds, their extinction and the rife of others in their place, Mr. Darwin again fays: "Through the procefs of fubftitution the old Englifh hound has been loft; and fo it has been with the Irifh wolf-dog, the old Englifh bulldog, and feveral other breeds, fuch as the alaunt, as I am informed by Mr. Jeffe. But the extinction of former breeds is apparently aided by another caufe; for whenever a breed is kept in fcanty numbers, as at prefent with the bloodhound, it is reared with fome difficulty, apparently from the evil effects of long-continued clofe interbreeding." [2] Many an ex-tinct breed (unlefs the animals exifted only in the imagination of their painters) may be feen in Berjeau's illuftrations of dogs, taken from old fculptures and pictures. And every admirer of Dürer's pictures muft remember the curious hairy dog with large ears, fomething like an eccentric Scotch terrier, which appears in fo much of his work; while at other times a dog is introduced which refembles a modern bull-terrier pup, both of which, however, it would be difficult to find examples of at the prefent day.

Mr. J. E. Harting confiders that all the dif-ferent breeds of our dogs may be conveniently deduced from the croffing of fix large groups:

[1] "Varieties of Plants and Animals under Domeftication," i., p. 44.   [2] *Ibid.*, i., p. 45.

1, the wolf-like dogs; 2, greyhounds; 3, fpaniels; 4, hounds; 5, maftiffs; 6, terriers. Profeffor Fitzinger enumerates more than 180 kinds of domeftic dogs. Mr. Harting alfo notes that all the dogs of Gaul and ancient Britain had erect or femi-erect ears, like wild dogs.[1]

A very important notice of Britifh dogs, to continue our chronological furvey, is recorded by Strabo, a contemporary of Cæfar. After fpeaking, like the latter, of the herds[2] of cattle to be feen in Britain, he adds that "hides, flaves, and *dogs of good breeding ufeful for hunting* are exported from it. The Kelts alfo ufe both thefe and the dogs of their own lands for warlike purpofes."[3] Thus the geographer curioufly enough comprifes Britifh dogs under the fame two heads as, it has been feen, they are arranged by the early ceramic arts of Britain. Pliny tells us that the Britons were wont to breed their dogs from wolves.

The next citation demands a long leap, to Oppian's time, A.D. 140. Here we firft meet with the term *agaffeus*, which has been fo varioufly interpreted. It is often rendered "beagle," and by fome "gazehound," which feems to mean a large hound running by fight, like the Irifh hound, or the prefent Scotch deerhound. And fo Tickell writes:

---

[1] Davis Lecture, July 3, 1884.

[2] Compare, too, Eumenius, "Panegyric of Britain,"— "tanto læta munere paftionum."

[3] κύνες εὐφυεῖς πρὸς τὰς κυνεγεσίας, Κελτοὶ δὲ καὶ πρὸς τοὺς πολέμους χρῶνται καὶ τούτοις, κ. τ. λ. (See "Monumenta Historica Britannica," 1848, vol. i., p. 141.)

" See'ſt thou the gazehound ? how, with glance ſevere,
From the cloſe herd he marks the deſtined deer ?"

To our mind, however, Oppian's deſcription appears to apply to no Britiſh dog ſo well as to a
Scotch terrier. We ſubjoin a tranſlation of his
ſonorous Greek hexameters :

" There is a certain kind of whelps apt for
tracking game, but of ſmall power ; little in ſize,
but worthy of much ſong, theſe the fierce tribes
of painted Britons rear, and they are known particularly as *agaſſæi.* In point of ſize they reſemble thoſe good-for-nothing dainty houſehold
pets, lapdogs ; round in ſhape, with very little
fleſh on their bones, covered with ſhaggy hair,
ſlow of viſion, but armed on their feet with cruel
claws, and ſharply provided with many poiſonous
canine teeth. For its ſcenting powers, however, the
agaſſeus is chiefly renowned, and it is excellent at
tracking, ſince it is very ſkilled to diſcover the
leaſt footprint of any running game, and even to
mark the very taint of its quarry in the air." [1]

Again the poverty of the times in literature
compels us to leap over rather more than a
century to Nemeſianus. This Carthaginian poet
alſo celebrates the hunting-dogs of Britain :

" Sed non Spartanos tantum tantumve Molossos
Paſcendum catulos, diviſa Britannia mittit
Veloces, noſtrique orbis venatibus aptos." [2]

We have another ſcrap relating to Britiſh dogs

[1] Oppian, " Cyneg.," i. 468. This deſcription in the original
is a very favourable ſpecimen of Oppian's ſtyle.
[2] Nemeſiani, " Cyneget," v. 123.

in Claudian (about A.D. 400). He fpeaks of the moloſſus " hunting with tender noſe ;" and again, of the " immortal moloſſus barking amid the thick miſts ſurrounding the mountain-tops,"[1] which are probably not maſtiffs in general (or from the context Britain might perhaps claim them), but ſtrictly the dogs of the tribe Moloſſi. Soon afterwards, amid an enumeration of different dogs, he does ſpecify the Britiſh maſtiffs:

" Magnaque taurorum fracturæ colla Britannæ."

From theſe ſemi-claſſical notices the antiquarian ſtudent of Engliſh dogs will not find much to detain him till he comes to the early Foreſt Codes. Thus Cnut's " Foreſt Laws,"[2] in Canon 31, lay down that " no man of mean eſtate ſhall have or keep the dogs called by the Engliſh ' greyhounds.' A freeman may, provided that their expeditation ſhall have been effected in the preſence of the chief foreſter."

Again, Canon 32 (tranſlated by Manwood), allows " thoſe little dogges called Velteres, and ſuch as are called Ram-hundt (al which dogges are to ſit in one's lap), may be kept in the foreſt, becauſe in them there is no daunger, and therefore they ſhall not be hoxed or have their knees cut."

As another ſpecimen of the ferocity of the ancient foreſt laws of our early kings, the following may be adduced: Canon 34, " If any mad dog

---

[1] " De Cons. Stilich.," iii. 294.
[2] "Ancient Laws of England," publiſhed by the Record Commiſſion, 1841.
[3] Manwood's " Foreſt Lawes," 1615.

ſhall have bitten a wild beaſt, then he ſhall make amends according to the value of a freeman, which is twelve hundred ſhillings. If, however, a royal beaſt ſhall have been killed by his bite, he ſhall be guilty of the greateſt crime."

Much that is intereſting connected with dogs uſed for falconry and the chaſe may be found in the "Boke of St. Alban's," 1486; but no Engliſh writer treated ſyſtematically of the different breeds of Britiſh dogs until John Caius, or Kayes, wrote his celebrated tractate "Of Engliſhe Dogges, the diverſities, the names, the natures, and the properties." Having been addreſſed in Latin to the famous Conrad Geſner, in order to aid that naturaliſt in his hiſtory of animals, it was tranſlated into Engliſh by "Abraham Fleming, Student," with the motto, "*Natura etiam in brutis vim oſtendit ſuam*," and publiſhed in 1576.[1] A highly euphuiſtical dedication to his patron, the Dean of Ely, was prefixed by this ſame Fleming, who alſo perpetrated ſome verſes on dogs on the reverſe of the title-page, entitled "A Proſopopoicall ſpeache of the Booke," which from their ſtyle and ſubject may moſt truly be termed one of the earlieſt ſpecimens of doggrel.

One or two intereſting facts attach to John Caius beſides the authorſhip of the earlieſt book on Engliſh dogs. This "jewel and glory of Cambridge," as Fleming ſtyles him, was born in 1510, and roſe to be a diſtinguiſhed phyſician.

---

[1] This has been reproduced in 1880 in a very convenient little volume (only changing the old Engliſh black-letter of the original into ordinary Roman type) at the *Bazaar* Office.

His name is ftill perpetuated in Gonville and
Caius College at Cambridge, which, after its firft
foundation by Edmund de Gonville in 1348, was
refounded by Caius, to whom it owes even more
than to its original founder.   A great portion of
the exifting College was built by Caius, and he
was for many years firft Fellow and then Mafter
of it.   Caius College is ftill the medical College
of the Univerfity, and can in paft years reckon
many notable phyficians amongft its fons, efpecially
Harvey, the difcoverer of the circulation of the
blood.   Perhaps even more honourable than this
is the diftinction Caius has obtained of being
alluded to in no obfcure manner by Shakefpeare.
" Mafter Doctor Caius, the renowned French
Phyfician," is one of the characters in "The Merry
Wives of Windfor" (1602); his fervants are Mrs.
Quickly and Rugby, while, characteriftically
enough, when angry with Sir Hugh, Shakefpeare
makes him fay, " By gar, he fhall not have a ftone
to throw at his *dog*" (" Merry Wives of Windfor,"
I., iv. 119).   Here it may be remarked incidentally that Shakefpeare, like the Bible, never fays a
good word for the dog, in fpite of its fidelity and
ufefulnefs.

The many divifions of his fubject which " that
prodigy of general erudition" (as Hallam calls
Gefner) was accuftomed to make, doubtlefs caufed
the plan to find favour in the eyes of his difciple,
Caius.   As the archæology of the dog ends with
his book, it is worth while giving an account of
it for the benefit of thofe dog-lovers who have

not yet made the acquaintance of this "breviary of Engliſhe dogges," as the author terms it.   His deſign is to "expreſſe and declare in due order, the grand and generall kinde of Engliſh Dogges, the difference of them, the uſe, the propertyes, and the diverſe natures of the ſame."   The treatiſe is eſpecially valuable for giving us the chief kinds of dogs then known in England (from which the pointer, it will be noticed, is abſent); but there are many quaint remarks and ſingular opinions alſo compriſed in it.   Firſt of all, Caius makes three great diviſions of the Engliſh dog :

> "A gentle kind, ſerving the game, [*i.e.* a well-bred kind].
> A homely kind, apt for ſundry neceſſary uſes.
> A curriſhe kind, meete for many toyes."

Theſe are ſubjeſted to ſundry more careful diviſions; and, finally, the firſt claſs is ſubdivided into dogs for the chaſe and dogs uſeful in fowling, under which heads the animals themſelves are one by one particularly deſcribed.

Of dogs uſeful in the chaſe, Caius enumerates "Hariers, Terrars, Bloudhounds, Gaſehounds, Grehounds, Leviners or Lyemmers, Tumblers, Stealers."   The harrier is our modern hound; and, if the author's claſſification of its duties may be truſted, was put in his day to very miſcellaneous uſes.   It has "bagging lips, and hanging eares, reachyng downe both fydes of their chappes," and was uſeful to hunt "the hare, foxe, wolfe, harte, bucke, badger, otter, polcat, lobſter (! !), weaſell,

and conny "—only " the conny," Dr. Caius
explains, " wee ufe not to hunt, but rather to take
it, fomtime with the nette, fometime with the
ferret." The terrar " creepes into the grounde,
and by that meanes makes afrayde, nyppes and
bytes the fox and the badger." It is evidently
the original of the modern fox-terrier. On the
bloodhound the author enlarges with evident
delight. It is ufeful, he fays, to track wounded
deer or their poachers, and is kept " in clofe and
darke channels " (kennels) in the day-time by its
owner, but let loofe at night, " to the intent that
it myght with more courage and boldneffe practife
to follow the fellon in the evening and folitary
houres of darkneffe, when fuch yll-difpofed varlots
are principally purpofed to play theyr impudent
pageants and imprudent pranckes." Thefe hounds
are alfo much ufed, he tells us, on the Borders
againft cattle-lifters. The females are called
braches, in common with " all bytches belonging
to the hunting kinde of dogges " (conf. Hotfpur's
words, 1 Henry IV., iii. 1, " I had rather hear
Lady, my brach, howl in Irifh "). The gazehound (*agaffeus*) he defcribes as a northern hound,
which, " by the fteadfaftnes of the eye," marks
out and runs down any quarry which it once
feparates from the herd. It clearly in this place
refembles the prefent Scotch deerhound. The
"grehounde" is "a fpare and bare kinde of dogge,
of flefhe but not of bone ; and the nature of thefe
dogges I find to be wonderful by y' teftimoniall
of hiftories," for which he cites Froiffart. At the

preſent day greyhounds are generally ſuppoſed to
be remarkably lacking in any other virtue than
that of ſpeed; all other points in their breeding
are neglected to enſure this good quality. The
lymmer (from *ligo*, becauſe held in a leaſh) is "in
ſmelling ſingular, and in ſwifteneſſe incomparable."
It is little uſed in England at preſent, but may be
ſeen in Brittany and on the Continent, where it is
a uſeful creature in the miſcellaneous collection of
big hounds employed to hunt the wolf and boar.
The *vertagus*, or tumbler, is another dog little
known in England now. It was wont to friſk
and tumble over and over, and by its antics
faſcinated rabbits and the like, until, gradually
drawing nearer, it made a ruſh at them. It
ſurvives in the little dog employed by the few
fowlers in the fens which yet exiſt, in order to
lure the wild-fowl, who have been attracted by
the decoy-ducks, further into the "pipe" of the
net. "The dogge called the theeviſhe dogge"
finds its modern exemplification in the "lurcher"
of gipſies and poachers. "At the bydding and
mandate of his maſter it ſteereth and leereth
abroade in the night, hunting connyes by the ayre
which is levened with their ſaver, and conveyed to
the ſenſe of ſmelling by the meanes of the winde
blowing towardes him. During all which ſpace of
his hunting he will not barcke, leaſt he ſhould bee
preivdiciall to his owne advantage."

Fowling dogs are the ſetter, the water-ſpaniel,
and "the dogge called the fiſher, in Latine *canis
piſcator*." Dr. Caius here ſomewhat unconſciouſly

imitates the famous chapter " Concerning Snakes
in Iceland," for he is fain to confeſs, in his chapter
on the " Fiſher," that " aſſuredly I know none of
that kinde in Englande, neither have I received by
reporte that there is any ſuche."   He appears to
confuſe it with the beaver or otter, and writes as
if the beaver were not yet extinct in England.
The whole chapter reminds an angler of the
celebrated queſtion which is raiſed in Walton's
book, whether the otter be beaſt or fiſh, ſolved by
the huntſman, who avows that, at any rate, " moſt
agree that her tail is fiſh."[1]   Indeed, the author's
wonderful diviſions of his ſubject irreſiſtibly ſuggeſt
that Shakeſpeare had this book in his mind when
he wrote :

> " Ay, in the catalogue ye go for men,
>   As hounds and greyhounds, mungrels, ſpaniels, curs,
>   Shoughs, water-rugs and demi-wolves are cleped
>   All by the name of dogs ; the valued file
>   Diſtinguiſhes the ſwift, the ſlow, the ſubtle,
>   The houſe-keeper, the hunter ; every one
>   According to the gift which bounteous nature
>   Hath in him cloſed ; whereby he does receive
>   Particular addition from the bill
>   That writes them all alike."[2]

Next our author comes to " the delicate, neate,
and pretty kind of dogges, called the ſpaniel
gentle, or the comforter, in Latine Melitæus or
Fotor " (from Melita or Malta, ſo anſwering to
our Malteſe dog).   Dr. Caius had evidently no
affection for theſe, and delivers himſelf of ſeveral
cauſtic ſentences, which may well be quoted for

[1] " Compleat Angler," i. 2.
[2] " Macbeth," iii. 2 (written in 1606).

the benefit of a good many "filly women" at
prefent : " Thefe dogges are litle, pretty, proper,
and fyne, and fought for to fatiffie the delicate-
neffe of daintie dames and wanton womens wills
inftrumentes of folly for them to play and dally
withall, to tryfle away the treafure of time, to
withdraw their mindes from more commendable
exercifes, and to content their corrupted con-
cupifcences with vaine difport" (a felly fhift to
fhunne yrckfome ydleneffe)." And again, " that
plaufible proverbe verified upon a Tyraunt,
namely that he loved his fowe better than his
fonne, · may well be applyed to thefe kinde of
people who delight more in dogges that are
deprived of all poffibility of reafon, than they doe
in children that be capeable of wifedome and
judgement."

Another chapter leads to the *canes ruftici*—the
dogs properly affociated by the ancients with
Great Britain. And firft comes the fhepherd-dog,
which, the author explains, need not be fierce, as,
thanks to King Edgar, England holds no wolves.
The maftiff, or bandog, which " is vafte, huge,
ftubborne, ougly and eager, of a hevy and
burthenous body, and therefore but of litle
fwiftneffe, terrible, and frightfull to beholde, and
more fearce and fell than any Arcadian curre
(notwithftanding they are faid to have the generation of the violent lion)," obtains a long notice with
divers hiftorical anecdotes. A good many crofs-
divifions follow in as many different fections treating of the " dogge-keeper " (or watch-dog) ; the

butcher's dog; the Moloffus; the dog that carries
letters and the like wrapped up in his collar; the
" mooner, becaufe/ he doth nothing elfe but watch
and warde at an ynche, wafting the wearifome
night feafon, without flumbering or fleeping, baw-
ing and wawing at the moone, a qualitie in mine
opinion ftraunge to confider;" the dog that draws
water out of wells; and the " Tyncker's curre,"
which many can yet remember drawing pots and
kettles about the country. Moft of thefe, adds
the author, are excellent dogs to defend their
mafter's property; and fome are very " deadly, for
they flye upon a man, without utterance of voice,
fnatch at him, and catche him by the throate, and
moft cruelly byte out colloppes of fleafhe."

The next chapter contains an account of "curres
of the mungrell and rafcall fort," which may be
called " waps " or warners. The turnfpit and
dancer (fo called becaufe taught to dance and
perform antics for gain) are treated of herein. It
would be unlike the author's age to forget the
marvels of canine life, fo his book concludes with
a chapter " of other dogges wonderfully engendered
within the coaftes of this country; the firft bred
of a bytch and a wolf (*lycifcus*); the fecond of a
bytyche and a foxe (*lacæna*); the third of a beare
and a bandogge (*urcanus*)." A few clofing words
are entitled, "a ftarte to outlandifh dogges," which
bear hardly upon Scotch and Skye terriers, now fo
common as pets, fo ufeful, and, it may be added,
fo faithful. Like Dr. Johnfon, Caius evidently
could not abide anything Scotch. " A beggerly

beaſt brought out of barbarous borders, fro' the uttermoſt countryes Northward, etc., we ſtare at, we gaſe at, we muſe, we marvaile at, like an aſſe of Cumanum, like Thales with the braſen ſhancks, like the man in the Moone." And ſo we heartily bid farewell to Dr. Caius and his amuſing tractate, ſtuffed full ("farſed" he would term it) of quaint ſentiments and recondite alluſions. It is a book which will delight all dog-lovers, independently of its value in continuing the hiſtory of their favourite animal from claſſical times. Perhaps it is worth adding that he repeats the old receipt for quieting a fierce dog which attacks a paſſer-by, viz., to ſit down on the ground and fearleſſly await his approach. Whether anyone has ever tried to put it in practice in real life we know not, nor have we ever cared to eſſay its virtues; but Ulyſſes certainly knew its value, and tried it to ſome purpoſe (ſee Plin., "Nat. Hiſt.," viii. 40; and "Odyſſey," xiv. 31).

Chaucer, like Shakeſpeare, ſeems to have had no great affection for dogs, but has not forgotten them in his portrait of the Prioreſs, Madam Eglantine. Her humanity and tenderneſs had to be deſcribed, and her love for her dogs gave the needful opportunity.

> "Of ſmale houndes hadde ſhe, that ſhe fedde,
> With roſted fleſh, and milk, and waſtel brede,
> But ſore wept ſhe if on of hem were dede,
> Or if men ſmote it with a yerde ſmert,
> And all was conſcience and tendre herte." (*Prologue.*)

In the ſtory of "The Pardonere and Tapſtere," another kind of dog is deſcribed:

" A whelp
That ley undir a fteyir, a grete Walfh dog,
That bare about his neck a grete huge clog,
Becaufe that he was fpetoufe, and wold fone bite."

Though the poems of Tickell and Somerville can fcarcely, in point of time, be deemed old enough to merit an antiquarian's notice, yet are they fufficiently remote from the prefent generation's reading to warrant here a word or two, which may aptly conclude thefe notes. A fragment of a poem on hunting by the former, the friend and mourner of Addison, is marked with all his claffic eafe and grace. The following lines will illuftrate at leaft one of Dr. Caius's dogs. Tickell bids his reader mark :

" How every nerve the greyhound's ftretch difplays,
The hare preventing in her airy maze ;
The lucklefs prey how treach'rous tumblers gain,
And dauntlefs wolf-dogs fhake the lion's mane ;
O'er all the bloodhound boafts fuperior fkill,
To fcent, to view, to turn and boldly kill."

And what reminifcences of the "Georgics" breathe in this portrait of a hound! We truft thefe famples may induce fome readers to turn to a poet who has been too long unjuftly neglected :

" Such be the dog I charge, thou mean'ft to train,
His back is crooked, and his belly plain,
Of fillet ftretch'd and huge of haunch behind,
A tapering tail that nimbly cuts the wind ;
Trufs-thighed, ftraight-hamm'd, and fox-like form'd his paw,
Large legged, dry foled, and of protended claw ;
His flat wide noftrils fnuff the favoury fteam,
And from his eyes he fhoots pernicious gleam,
Middling his head and prone to earth his view,
With ears and cheft that dafh the morning dew :

He beft to ftem the flood, to leap the bound,
And charm the Dryads with his voice profound ;
To pay large tribute to his weary lord,
And crown the fylvan hero's plenteous board."

Gervafe Markham's quaint picture of the "water
dogge" may well be compared with this (fee his
"Hunger's Prevention," London, 1621, in which
are a good many more notices of dogs): "His
Necke would bee thicke and fhort, his Breft like
the breft of a fhippe, fharp and compaffe; his
Shoulders broad, his fore Legs ftraight, his chine
fquare, his Buttockes rounded, his Ribbes compaffe, his Belly gaunt, his Thyes brawn, his
Gambril crooked, his pofteriors ftrong and dewe
clawde, and all his four feete fpacious, full and
round, and clofed together like a water duck"
(chap. ix.).

Much curious matter on dogs may be picked
out of George Turberville's "Book of Faulconrie,"
publifhed in 1575; and his "Noble Arte of
Venerie," in which he largely compiled from Du
Fouilloux and Jean de Clamorgan. Harington,
Glanville, Barlow, and William Harrifon, in
Holinfhed's "Hiftory" ed. 1586, cap. 7, may
alfo be confulted with profit. Some of this old-
world learning has been brought together by Mr.
G. R. Jeffe in his "Refearches into the Hiftory of
the Britifh Dog" (London, 1866). All thefe
authors love dogs as fervently as the Indian hero,
Yoodhift'huru. When the chariot of Indru was
waiting to convey him to heaven, he came attended
by his dog. "I don't take dogs," faid Indru.
"Then I don't go," replied Yoodhift'huru. The

dog, however, turns out to be Humu, a god, and the difficulty was got over (fee Berjeau's "Varieties of Dogs in old Sculptures," etc., London, 1863, p. 1).

Somerville's four books in blank verfe on the chafe are, perhaps, too lengthy for readers who tire quickly of Milton; but the adventurous explorer will find fome landfcapes in them which betray no mean defcriptive fkill, lit up every here and there by a flafh of imagination. He, too, was evidently a dog-lover ; and feveral good defcriptions of the hounds which found favour with huntfmen at the beginning of the laft century atteft his enthufiafm for hunting. After his verfes no further excufe can be found for continuing the fubject, though it is worth while to add that a few notices on dogs are contained in Pepys's " Diary."

CHAPTER IV.

THE CAT.

FAMILIAR to all as is the domeſtic cat, a number of intereſting queſtions are involved in its early hiſtory. A diſtinguiſhed biologiſt has recently taken it as the type of the *felidæ*, and filled a goodly volume on it without by any means exhauſting the ſubject.[1] The origin of the large family of cats, both foſſil and living ſpecies, is traced in geologic time by Lyell and Owen to the Pliocene Period, when, together with the *canidæ*, cats alſo came into being. Profeſſor Owen enumerates as foſſil ſpecies *F. Spelæa*, great cave tiger, whoſe remains have been found in Kent's Hole and elſewhere ; *F. Pardoides*, of which one tooth was found by Mr. Lyell in the Red Crag, Newbourn, in 1839 ; *F. Catus*, the wild cat, probably identical with the preſent wild cat of the north ; and a huge ſabre-toothed feline animal as large as a tiger, and, to judge from its teeth, more

[1] St. John Mivart's " The Cat " (Murray, 1881).

deſtructive, *Machærodus latidens.* Its remains have alſo been found in Kent's Hole and at Kirkdale.[1] Mr. Mivart, however, who has more recently inveſtigated the ſubject, enumerates, with deſcriptions, fifty diſtinct ſpecies of living cats, and adds, " A much larger number of ſpecies have probably exiſted in the paſt." The great cat, known as the cave lion (*F. Spelæa*), lived in England in middle and late pleiſtocene times; but Mr. Mivart traces the anceſtry of cats to a much more diſtant period. " The remains of certain large cats have been found in pliocene, and miocene, and even in eocene depoſits, which differ from any exiſting cats in the enormous size of their upper canine teeth, *e.g., machærodus, hoplophoneus, pseudælurus,*"[2] etc. There are ſigns that the cat was domeſticated in the bronze period.

It is commonly ſuppoſed that the wild cat is the anceſtor of our domeſtic cats, but this is certainly a miſtake.[3] Few animals are more irreclaimable than the wild cat. One which the Duke of Sutherland, as head of the Clan Chattan, or Clan of the Cats, exhibited in a ſtrong cage at the Cryſtal Palace ſome years ago during a ſhow of cats, flew fiercely at all who approached it. No amount of kindneſs appears to tame it; and the progeny invariably revert to a wild life in the

[1] Owen, "Hiſtory of Britiſh Foſſil Mammals and Birds." (Van Voorſt, 1846), p. 173.

[2] Mivart, *ut ſup.*, pp. 431, 432.

[3] Prof. Owen thinks that " our houſehold cat is probably a domeſticated variety of the ſame ſpecies which was contemporary with the ſpelæan bear, hyæna and tiger." (" Britiſh Foſſil Mammals," p. 173.)

woods as foon as poffible. Befides which, the period of geftation of the wild cat is fixty-eight days, twelve days longer than that of the domeftic animal.[1] The late Mr. F. Buckland, too, pointed out, as a ftriking difference between it and the domeftic animal, that its inteftines are much shorter than thofe of the latter animal. Thus they were found to be only five feet in two fpecimens of the wild cat, whereas they would probably be three times that length in the domefticated creature.[2] This ftatement, however, requires confirmation. In the spring of 1884 a fuppofed wild cat was fhot in a large extent of woodland near Wragby, in Lincolnfhire, called Bullington Wood. Wild cats are fuppofed to have been extinct in this county from quite the beginning of the century. This cat was ftuffed, and feen by many others as well as the writer. It might either be a true wild cat—in which cafe it had efcaped from confine-ment—or elfe was a furvival of the true old Britifh wild cat. It is curious, in connection with this, that the laft locality in which the kite was feen in this country was in thefe very woodlands; and the marten is yet found there. It feemed the general opinion that this was a true wild cat; with the writer, however, another alternative found favour—that it was a defcendant, perhaps in the fourth or fifth generation, from an efcaped domeftic cat. It is a fingular fact that efcaped cats and

[1] Mivart, *ut fup.*, pp. 2-6.
[2] "Logbook of a Fifherman" (1875), p. 252. Darwin, "Plants and Animals under Domeftication," ii. 292.

their progeny have much tendency to revert in colour and appearance to the type of the true *f. catus*. Its colour is a dark-grey, or grey-brown ftriped with black. The cat in queftion seemed to the writer too rich in colour, with an under-fhade of yellow, which was fufpicious. Colour, however, is proverbially deceitful in natural hiftory inveftigations. The head was too round, the legs too flender, and the tail not fufficiently abrupt; and thefe are important ftructural differences. *Adhuc fub judice lis eft*. Two or three other hints, moreover, feem to point to the conclufion that the domeftic cat is a foreign importation. The curious penalty, for inftance, denounced in the old Welfh laws againft him who fhould kill the king's cat, "the keeper of the royal granary," appears to fuggeft that a cat was a fomewhat rare and valuable animal. The offender was compelled to pay as much corn as would cover the cat's body when held up by the tip of its tail. Dick Whittington and his cat is another indication of the foreign extraction of the animal. Being fent to Barbary, it fold for a good price, and enriched its mafter.

All the evidence points to Egypt as the country where cats were originally domefticated in the Weft, though it was known in India 2,000 years ago. They are wrong who derive the cat's appearance in Europe from Perfia, and ftate that its name Pufs is a mere diminutive of Perfe. Dr. Brugfch-Bey fhews that one of the titles of Ofiris was Bafs, the cat (or leopard), whence, with more probability, comes our word Pufs. His wife, Baft (the

"biſſat" or tabby cat of the modern Arabians),
gave her name to Bubaſtis (Pi-Baſt, the City of
Baſt).[1] The parent of our cat is to be ſought,
either in the *felis bubaſtes* or the *f. caligata*
(*maniculata*), found at preſent wild in Egypt.
Probably the latter, with an admixture of other
ſtrains, is the original ſtock. It is a native of
Northern Africa, about a third ſmaller than our
wild cat, and of a yellowiſh colour, ſomewhat
darker on the back and whitiſh on the belly.
Thus Egypt, the granary of the ancient world,
naturally was the firſt country of the Weſtern
world to domeſticate the cat. It is mentioned in
inſcriptions as early as 1684 B.C., and was certainly kept as a pet in Egypt 1,300 years B.C.
The earlieſt known repreſentation of the cat as a
domeſtic creature is on a tablet of the eighteenth
or nineteenth dynaſty at Leyden, wherein it appears
ſeated under a chair. It was venerated in certain
diſtricts of ancient Egypt:

" Illic *æluros*, hic piſcem fluminis, illic
Oppida tota canem venerantur, nemo Dianam."
(Juv., *Sat.* 15, 7.)

---

[1] See Burton, "Land of Midian" (Kegan Paul and Co.,
1879), vol. i., p. 113. The above learned Egyptologiſt would
derive Bacchus and his prieſts, the Bacchi and Bacchantes, from
the Oſiric term, Baſs. It is at leaſt a curious fact that the dreſs
of theſe prieſts conſiſted of a leopard's ſkin.

"According to Lenormant, the cat does not appear on
Egyptian ſculpture earlier than the thirteenth dynaſty (2020,
B.C.), and therefore the credit of its domeſtication is due to the
inhabitants of the Upper Nile. This proceſs, remarks Hehn,
muſt have taken a long time, but it was thoroughly ſucceſſful
in the end." (W. R. S. Ralſton, *Nineteenth Century*, Jan.,
1883.)

The goddefs Pafht or Bubaftis, the goddefs of cats, was under the Roman Empire reprefented with a cat's head, that creature being efteemed an emblem both of the fun and the moon by the ancient Egyptians, partly from its eyes being fuppofed to vary with the courfe of the fun, partly becaufe they were thought to wax and wane with the moon. Dr. Birch ftates that the earlieft reprefentation of the cat with which he is acquainted and of whofe date he is certain, is to be found on a tomb in the Berlin Mufeum, apparently of about 1600 B.C. It alfo appears in hunting-fcenes of the eighteenth dynafty, and in rituals written under that dynafty, but probably repetitions of a much earlier text. At times it is in a boat with the hunters, but eager to be allowed to fpring into the thickets of aquatic plants ; and again it is reprefented among the birds ftruck down by the fowler, and apparently taught to work either as a fpringer of the game or as a retriever. When the facred cats died, their bodies were always embalmed, and behind a temple at Beni Haffan, dedicated to Bubaftis, are pits containing a multitude of cat mummies.[1] When Herodotus vifited Egypt, he was naturally ftruck with the exaggerated reverence paid to cats, and devotes a quaint chapter to them which is well worth tranflating. Two facts come out in it ; firft, a certain fcarcity of cats even in Egypt ; and fecondly, the facrednefs of the animal.

[1] Mivart, *ut sup.*; and Wilkinson, "Ancient Egyptians," i., p. 236.

" Though the Egyptians have many domeſtic animals, there would be many more did not the following circumſtances occur. When kittens are born, their mothers are unwilling to conſort with the males, ſo the Toms have deviſed a plan to remedy this. They carry off and kill the kittens ; but though they kill, they do not eat them. Then the mothers, having loſt their kittens, naturally long for others (the cat being an animal fond of young ones), and ſo again ſeek the males. When a fire breaks out, a divine impulſe comes over the cats. The Egyptians ſeparate and keep watch over them, negleĉting to put out the conflagration ; but the cats, ſlipping under and leaping over the men, ſpring into the fire. When this happens great grief takes poſſeſſion of the Egyptians, and wherever cats have thus periſhed of their own accord, all the inmates of the houſe ſhave off their eyebrows only ; but whenever a dog has died, their whole body and head. After their death, cats are borne off into ſacred abodes, where, after having been made into mummies, they are buried in the City Bubaſtis" (Book ii. 66). Diodorus ſays that he ſaw the Egyptians murder a Roman who had accidentally killed a cat.

Chabas ſays that cats are not ſeen on any of the hieroglyphic tables illuſtrating the life of the Egyptians, but are often employed as the equivalent for the found "meou." The cat dates from the moſt ancient times in that country, and is mixed up with the oldeſt legends. This ſhews why it was frequently made into a mummy. It probably had a myſtical ſignificance, for—"dans quelques-unes des peintures parvenues juſqu'à nous, les anciens Egyptiens ſe montrent accompagnés de leurs chiens et de leurs ſinges favoris, auxquels ils donnaient des noms comme on le fait aujourd'hui ; le chat n'y figure jamais." The camel, again, is never repreſented on any of the ſurviving monuments, yet it was known to the Egyptians in the time of Abraham (Chabas, "Études," pp. 406, 408 : Paris, 1873).

Cats and hares ſhare an equal notoriety in the annals of witchcraft. "When one of us" (ſays one of the Culdean witches) "is in the ſhape of a cat, and meet with any others of our neighbours, we will ſay, 'Devil ſpeed thee, go thou with me,' and immediately they will turn to the ſhape of a cat and go with us." There was a large aſſembly and fight with ſuch cats at Scrabſter, in the north of Scotland, 1718. The marvellous recital tells how one Mr. William Montgomery valorouſly ſtuck one with his dirk through the hinder quarters to a cheſt, "yet after all ſhe eſcaped out of the cheſt with the dirk in her hinder quarters" (J. H. Burton, "Criminal Trials in Scotland," vol. i., p. 290 : London, 1852). Freja, in the "Northern Mythology," rides to the battlefield in a waggon drawn by two cats, this animal being ſacred to her. Hence it is popularly aſſigned to hags, witches, etc. When a bride goes to her wedding in fine weather the Germans ſay, "She has ſed the cat well ;" *i.e.*, not offended the favourites of the love-goddefs (Grimm's "Northern Mythology," tranflated by Stallybrafs, i., p. 305.)

In fpite of the proverb:

"Catus sæpe satur cum capto mure jocatur,"

Mr. St. John Mivart is of opinion that the cat, when tormenting a moufe, is not doing fo from native cruelty, but in order to keep her claws in order, juft as her big brother, the tiger, is compelled to fcratch the bark of trees, efpecially the Indian fig-tree, in order to cleanfe his claws. Japanefe cats, like thofe of the Ifle of Man, are taillefs. The cat is a favourite on tavern figns; our own Cat and Fiddle matching the Flemifh " Le Chat qui Fume," and the equally well-known " Chat de St. Jean " with its long tobacco-pipe.

Cats were not domeftic animals with the Hebrews, any more than dogs. It is not furprifing, therefore, that they are paffed over in filence in Holy Scripture. In Baruch vi. 22, indeed, is a curious paffage which occurs in what purports to be a letter of Jeremiah to the captives about to be led into Babylon by Nebuchadnezzar. In it the prophet tells them of the fenfelefs idols they will there fee, and adds, " upon their bodies and heads fit bats, fwallows, and birds, and the cats alfo ;" but the paffage is in all probability a forgery of the firft century B.C.

From Egypt cats feem to have been introduced into Greece, and thence into Rome. A frefco painting of a cat was difcovered at Pompeii. Thefe animals were not much prized, however, by either Greeks or Romans. The only paffage in the claffics where the word which has paffed into our "cat" occurs is in an epigram of Martial (xiii. 69):

"Pannonicas nobis nunquam dedit Umbria *cattas*."

Phædrus has a fable of an eagle, a cat, and a fow which inhabited the top, middle, and bafe of an oak, and clearly ufes the word *felis* of our well-known cat.   One line exactly expreffes the cat's nocturnal habits:

"Evagata noctu suspenso pede." (*Fab.*, 2, 4.)

Compare, too, the proverb, "Felem Minervæ."

The connection of cats and Egypt comes out again in a paffage of Ovid ("Met.," v. 330).   A mufe fings:

"How the gods fled to Egypt's flimy foil,
    And hid their heads beneath the banks of Nile,
    How Typhon from the conquered fkies purfued
    Their routed godheads to the feven-mouthed flood ;
    Forced every god, his fury to efcape,
    Some beaftly form to take or earthly fhape ;
    Jove (fo fhe fung) was changed into a ram,
    From whence the horns of Libyan Ammon came ;
    Bacchus a goat, Apollo was a crow,
    Phœbe a cat, the wife of Jove a cow.'
                    (Maynwaring's Tranflation.)

The peculiar roughnefs of the tongue in the feline race generally is pointed out by Pliny (xi. 37, 65).   He adds: "With what filence, with what light footfteps do cats creep upon birds! how fuddenly, when they have fpied them, do they fpring out upon mice!" (x. 73, 202). Arguing from this and fimilar paffages, the late Prof. Rollefton and others believed that the domeftic animal of the Greeks and Romans, for which we now ufe the cat, was the white-breafted marten. The word *feles*, it is true, is commonly ufed for the weafel ; but, on the other hand, its Greek fynonym αἰλουρος, according to the beft derivation by Buttemann, applies exactly to the wavy motion of the tail fo peculiar to the cat family. The Englifh term " cat " probably comes from the Latin *catus* (cunning). In Anglo-Saxon documents it is found with the fpelling " catt." " When Julius Cæfar landed here," fays Mivart (*ut fup.*, p. 2), " our forefts were plentifully fup-plied with cats, while probably not a fingle moufer exifted in any Britifh town or village." The wild cat is at prefent reftricted to the extreme north and north-weftern diftricts of Scotland, having become extinct in England, and never feemingly having exifted at all in Ireland. But in the Middle Ages it was common in the wilder parts of England, as its fur was commonly ufed to trim dreffes. John, Earl of Morton, in a charter granting immunities to free tenants outfide the Regard of the Foreft of Dartmoor, fays: " Quod capiant capreolam, vulpem, *cattum*, lupum, leporem, lutrum ubicunque illa invenerint extra reguardum forefte mee,"[1] as if the wild cat were not uncommon at the end of the twelfth century. Pope Gregory the Great had a tame cat, and cats were often inmates of nunneries in the Middle

[1] See Rowe's " Perambulation of Dartmoor " (1848), p. 263. The Charter is in the poffeffion of the Dean and Chapter of Exeter.

Ages. Its flefh was interdicted as food, having
been a favourite difh with the heathen Northmen.

A curious parallel to Whittington and his cat
occurs in a petition of the year 1621 of one William Bragge to "the Company of the Eaft India
and Sommer Iflands," claiming £6,875 for divers
fervices rendered.[1] Among their recital is found:
"Item, more for 20 Dogges and a greate many
Catts which, under God, as by your booke written
of late, ridd away and devoured all the Ratts in
that Iland [Bermuda], which formerly eate up all
your corne, and many other bleffed fruites which
that land afforded. Well, for theis, I will demand
of you but 5lb. a piece for the Doggs, and let the
Catts goe—100lb. os. od." Hone relates that on
the Feftival of Corpus Chrifti at Aix in Provence,
"The fineft Tom cat of the country, wrapped in
fwaddling clothes like a child, was exhibited in a
magnificent fhrine to public admiration. But at
the Feftival of St. John poor Tom's fate was
reverfed. A number of the tabby tribe were put
into a wicker-bafket, and thrown alive into the
midft of an immenfe fire, kindled in the public
fquare by the Bifhop and his clergy. Hymns and
anthems were sung, and proceffions made by the
people in honour of the facrifice."[2] It is fingular
to find thefe traditions of the facrednefs of the
animal lingering in Europe in the Middle Ages.

The cat is a celebrated animal in folk-lore and
proverbs.[3] Perhaps Fuller's faying is one of the

[1] *N. and Q.*, 3rd S., 2, 345.
[2] "Every Day Book," vol. i., p. 758.
[3] Darwin, "Origin of Species," p. 9, ed. 6.

moft ungallant of the latter : " A cat has nine lives, and a woman has nine cats' lives." Almoft equally paradoxical with this proverb appears at firft fight what is neverthelefs regarded as a true law of nature, that cats which are entirely white, and have blue eyes, are generally deaf ; but it has lately been stated that this peculiarity is confined to the males. " Care killed the cat " is another proverb which reflects upon the eafy lives led by thefe animals. The circumftances of their owners do not affect them, and a cat is a faturnine creature, equally happy and at home whatever befalls her mafter. The familiar prefence of the cat on every hearth comes out in " a cat may look at a king." Why Chefhire cats fhould always grin is fome-what infcrutable, but fo fays the Scotch proverb.[1] Shakefpeare was aware of the cat's weaknefs for fifh, but its unwillingnefs to wet its feet in catching them, and applies it finely. Lady Macbeth taunts her hufband when he hangs back from the murder with

> " Letting I dare not wait upon I would,
> Like the poor cat i' the adage,"

—referring to the mediæval adage,

> " Catus amat pifces fed non vult tingere plantas ;"

and the fame poet well knew the nature of the true wild cat :

> "He fleeps by day
> More than the wild cat."—(*Merch. of Venice*, ii. 5.)

---

[1] In Sicily the cat is facred to St. Martha. He who kills a cat will be unhappy for feven years. Europe has always regarded the cat as a diabolical creature. A Ruffian proverb fays that a black tom cat at the end of feven years turns into a devil. (Ralfton, *ut fup.*)

F

To fhow the manner in which one part of nature influences and acts upon others until the fauna or flora of a diftrict may be changed by what feem, taken feparately, infignificant caufes, it is worth while quoting a fpeculation of Darwin, in which the cat plays a confpicuous part. " The common red clover is only vifited by humble-bees, as hive-bees cannot reach the nectar. The heartfeafe (*viola tricolor*) is another plant which alfo feems to owe its fertilization only to humble-bees. It may be regarded, therefore, as highly probable that if the whole genus of humble-bees became extinct or very rare in England, the heartfeafe and red clover would become either very rare, or would altogether difappear. The number of humble-bees in a diftrict depends in a great meafure on the number of field-mice, which deftroy their combs and nefts. It is eftimated that more than two-thirds of them are thus deftroyed all over England. The number of mice is largely dependent upon the number of cats, and it has been found near villages and fmall towns the nefts of humble-bees are more numerous than elfewhere, which is attributed to the number of cats which deftroy the mice. Hence it is quite credible that the prefence of a feline animal in large numbers in a diftrict might determine, through the intervention firft of mice and then of bees, the frequency of certain flowers in that diftrict."[1]

" There remains to be told but one more cat ftory of importance. It claims to be of recent

[1] See "Origin of Species," *ut fup.*, p. 57.

date, and it conveys the uſeful moral that they who attempt to benefit their fellow-men muſt he prepared for diſappointment.   A few years ago, if newſpaper reports may be believed, a ſhip was ſent to the colony of Triſtan d'Acunha with a ſcore of cats on board.   Theſe animals were a preſent from the Lords of the Admiralty, to whom it had been reported that the iſland was mouſe-ridden.   When the veſſel arrived the Governor of the colony begged that the cats might be kept on board.   It was quite true, he explained, that the iſland was infeſted by mice, but it was alſo overrun by cats.   And in Triſtan d'Acunha, cats, in conſequence of ſome ſtrange climatic influence, always abandoned mouſing, a faét which accounted for the abnormal development of the mouſe population.   So that a gift of cats to Triſtan d'Acunha was even leſs likely to be welcome than a present of 'owls to Athens.' "[1]

The unhappy reader, however, will now turn upon the author with Bertram's words: " I could endure anything before but a cat, and now he's a cat to me!" (" All's Well that Ends Well," iv. 3, 265).

[1] From an admirable article by W. R. S. Ralſton (*Nineteenth Century*, Jan. 1883) on the folk-lore of cats, called " Puſs in Boots."

CHAPTER V.

OWLS.

THE difrepute into which owls have fo largely fallen with the ignorant appears to be due to the Romans, rather than the Greeks. In any dull country, indeed, where the nights are long and dark, the nocturnal cries and ftrange activity of the owl after dufk, its glaring eyes and frequently horned ears, will naturally imprefs the fuperftitious ; but what may be called its literary heritage of hatred and infamy comes to it from Italy. The owl in Homer is fimply " a long-winged bird," and appears in company with " falcons and chattering fea-crows, which have their bufinefs in the waters "[1] in the fair wood of alder and fweet-fmelling cyprefs which furrounded the pleafant cave of Calypfo. No ill-fame has yet attached itfelf to the bird. But reference to Pliny at once fhows the evil character it poffeffed at Rome, and gives the reafon for it. The city was indebted to

[1] " Odyssey," v. 66 (Butcher and Lang).

the Etrurians for its ſcience of augury, and it had
pleaſed the Etrurian *haruſpices* that the owl
ſhould be regarded as a bird of ill-omen. So
Pliny ſays: "The great horned owl is of mournful import, and more to be dreaded than all other
birds in auſpices connected with the ſtate. It in-
habits waſte places, and thoſe not merely deſerts,
but dreadful and inacceſſible localities ; being a
prodigy of night, making its voice heard in no
manner of ſong, but rather in groaning. So whenever ſeen in cities or in daylight it is a direful
portent. Perchance it is not ſo much fraught with
horror when ſeen ſitting on private houſes. It
never flies where it liſts, but is always borne along
in a ſlanting direction. Having once entered the
capitol, the city was purified on account of it in
the ſame year. There is an unlucky and incendiary bird, owing to which I find in the 'Annals'
that the city was repeatedly purified, as when
Caſſius and Marius were conſuls, in which year
alſo it was cleanſed, as a horned owl had been
ſeen. What this bird is I cannot find out, nor
does tradition tell. Some ſay that any bird is an
incendiary, if it appears bearing a coal from the
altars. Others call it a *ſpinturnix*" (*i.e.*, an
abominable bird), "but neither can I find anyone
to tell me what kind of bird this is. Another
confeſſion of general ignorance is that it was called
by the ancients 'a bird which forbade things to
be done.' Nigidius terms it a thieviſh bird, becauſe it breaks the eggs of eagles. There are,
beſides, ſeveral kinds, treated of in the Etrurian

ritual, which have now, marvelloufly enough, died out, although thofe birds which man's appetite lays wafte increafe. One Hylas wrote very fkilfully concerning omens, and tells that the owl, with feveral other predatory birds, comes out tail firft from the egg, inafmuch as the eggs are weighed down by the heavy heads of the chicks, and confequently prefent the tails of thefe birds to the cherifhing influence of the mother's body.

"Crafty is the mode in which owls fight other birds. When furrounded by a great number, they fling themfelves on their backs, and fight with beak and claws, their bodies being clofely contracted, and thus protected on all fides. The kite will help them, from a natural kinfhip in robbery, and fhares the combat. Nigidius fays that owls fleep for fixty days during winter, and have nine different cries."[1]

It is fmall wonder that if thefe were the kind of popular beliefs at Rome the unlucky owl obtained an ill-character in Latin, and tranfmitted the evil heritage to the Romance languages. Virgil, with his ftrong poetic feeling, introduces the bird fitly enough among the portents which prefaged the death of Dido, when abandoned by Æneas. "The lonely owl would frequently lament in funereal ftrains from the houfe-tops, and prolong her cries into a wail of woe" ("Æneid," iv. 462). Again, the fame poet fhews the triumph of good over evil when the return of fettled fine weather difcomforts the owl's melancholy prog-

[1] Pliny, x. 16-19.

noftics ; "in vain at fuch a time does the owl as fhe watches funfet from fome roof-top ply her ftrains of woe far into the night" ("Georg.," i. 402). In another poem he dwells upon the hoarfe notes of the owl as compared with the wild fwan's fonorous, mufical fong, "certent et cygnis ululæ" ("Ec.," viii. 56) ; the very name which he gives the unlucky bird expreffing its monotonous hootings. A common Greek name for the bird was "fcops," which alfo expreffes its hooting. The ordinary word for an owl in Greek, however, comes from the glaucous, or glaring character of the eyes in this bird. From the gleaming, flashing eyes which the poets attributed to Minerva, the owl became her bird, and is often reprefented in ancient art as her fymbol. The *strix pafferina* (*glaucidium pafferinum* of Linnæus) was thus regarded at Athens as the bird of wifdom, and from the abundance of owls at that city[1] arose the Greek proverb "owls to Athens," of fimilar meaning with our "coals to Newcaftle." The drachma, an Athenian coin, bore Minerva's head on one fide, and on the other an owl, and this device continued throughout the whole hiftory of the Athenian coinage. Naturally enough thefe coins were called "owls." The Greek tetradrachms alfo bore the imprefs of an owl, and, in the palmy days of Athens, had univerfal currency. Curioufly enough, Mr. R. F. Barton, among the coins which he discovered at

---

[1] *Athene noctua* and *Athene glaux* alfo owe their names to Athena and her city, Athens.

Maghair Shu'ayb (on the eaft of the Gulf of
Akaba), in his exploration of the land of Midian,
found that "the gem of his whole collection was a
copper coin thickly encrufted with filver, proving
that even in thofe days the Midianites produced
' fmafhers ;' fimilarly, the Egyptian miners ' did'
the Pharaoh by inferting lead into hollowed gold.
The obverfe fhews the owl in low relief, an animal
rude as any counterfeit prefentment of the Θεὰ
γλαυκῶπις 'Αθήνη ever found in Troy.   It has the
normal olive-branch, but without the terminating
crefcent (which, however, is not invariably prefent)
on the proper right, while the left fhews a poor
imitation of the legend AΘE(NH).   The filvering
of the reverfe has been fo corroded that no figns
of the goddefs' galeated head are vifible.   My
friend, Mr. W. E. Hayns, of the Numifmatic
Society, came to the conclufion that it is a barbaric
Midianitifh imitation of the Greek tetradrachm."[1]

The owl became in good truth a meffenger of
death to Herod Agrippa, who was fmitten of
God for not giving Him the glory, and died at
Cæfarea (Acts xii. 23).   " Prefently his flatterers
cried out," fays Jofephus,[2] " one from one place,
and another from another ; (though not for his
good), that ' He was a god ;' " and they added,
" Be thou merciful to us.   For although we have
hitherto known thee only as a man, yet fhall we
henceforth own thee as fuperior to mortal nature."

---

[1] "The Land of Midian" (Kegan Paul and Co., 1879).
Vol. i., p. 93.
[2] Whifton's Tranflation.   "*Antiq.*," Bk. xix. 8, § 2.

Upon this the King did neither rebuke them, nor rejeƈt their impious flattery. " But as he prefently afterward looked up, he faw an owl fitting on a certain rope, over his head ; and immediately un- derftood that this bird was the meffenger of ill tidings, as it had once been the meffenger of good tidings to him, and fell into the deepeft forrow." Severe pain at once came upon him, and he acknowledged that Providence was thus reproving the lying words which he had accepted from the people, and died five days afterwards. This paffage is alfo noticeable for a critical battle which has been fought over it ; as if Eufebius, the eccle- fiaftical hiftorian, had falfified thefe words of Jofephus to identify the owl with the angel of the Lord mentioned in the Book of Aƈts, the word " meffenger " in the above citation being in the original *angelus*, angel or meffenger. Whifton has a fatisfactory note on the point.

North America admires, but Arab folk-lore bears hardly upon the owl. Among the Red Indians the bird is believed to lament the golden age when men and animals lived in perfeƈt unity until it came to pafs that they began to quarrel, when the Great Spirit in difguft failed acrofs the feas, to return when they had made up their differences. So every night in the great pine forefts the fnowy owl repeats his· " Koo, koo fkoos!" " Oh, I am forry !" "Oh, I am forry !"[1] The fine owl of the Sinaitic Peninfula, however, is known by the

[1] Leith Adams's " Field and Foreft Rambles in New Bruns- wick," p. 58.

Arabs as " the Mother of Squeaking," and is believed to fuck out children's eyes. The owl and the hyena are ufed by the natives as charms ; the burnt feathers of the former, and the boiled flefh of the latter animal being confidered invaluable fpecifics for numerous diforders. In other parts of Arabia the hooting of the owl portends death, and the cry " Fât, fât " is interpreted " He is gone, gone !"[1] An owl appeared before the battle with the Parthians, in which Craffus fell, and was fuppofed by the ancients to prefage his death. Of all thefe beliefs old Sir T. Brown faid well, " which, though decrepit fuperftitions and fuch as had their nativity in times beyond all hiftory, are frefh in the obfervation of many heads, and by the credulous and feminine party ftill in fome majefty among us. And therefore the emblem of fuperftition was well fet out by Ripa in the picture of an owl, an hare, and an old woman."[2]

The difcuffions which have arifen from Dr. Schliemann's difcoveries of the fo-called owl pottery at Hiffarlik have been so frequently renewed of late that it is only neceffary to allude to them here. People had an opportunity of judging for themfelves in the exhibition of relics from old Troy at South Kenfington. Not uninftructive is a favourite Arabic apologue, though derived probably from the Perfian. The Saffanian King of Perfia, Bahram, was fo indifferent to the welfare of his fubjects that half the towns and villages in his

[1] Burton's " Land of Midian," vol. i. 142.
[2] " Vulgar Errors," v. 22.

kingdom became ruined and deferted. One night, while on a journey accompanied by a Mobed, or Magian prieft, he paffed through fome depopulated villages, and heard an owl fcreech, and its mate anfwer him. "What do the owls fay?" afked the King. The Mobed anfwered, "The male owl is making a propofal of marriage to the female, and the lady replies: 'I fhall be moft delighted, if you will give me the dowry I require.' 'And what is that?' fays the male owl. 'Twenty villages,' fays fhe, 'ruined in the reign of our moft gracious Sovereign Bahram.'" "And what did the male owl reply?" afked Bahram. "Oh, your Majefty!" anfwered the prieft. "He faid, 'That is very eafy; if his Majefty only lives long enough, I'll give you a thoufand.'" The leffon, fays hiftory, was not loft upon the King.

In French folk-lore the owl has acquired an evil name becaufe, when the wren had brought down fire from heaven, while the other birds in their gratitude contributed a feather apiece to replace its fcorched plumage, the owl refufed, alleging that fhe would require all her feathers during the approaching winter. On this account it has been condemned to eternal feclufion during the warm day, and to perpetual fuffering from cold during the night. This explains why "the owl, for all its feathers, was a' cold" on St. Agnes's Eve, and why the other birds pefter it if it appears in funfhine. An omelette made of owl's eggs is faid to be a cure for drunkennefs.

The poor bird, under its French name *effraie*, carries a continual remembrance of the old belief that it boded misfortune, *effraie* being a corruption of *frefaie*, which is connected with the Latin *præfaga*.[1] It is curious that the Hindoos make an owl fit upon the "inviolable tree" of their mythology (as if it were connected with life), near the tree which bears the *foma*, or drink of immortality. Returning once more to the Weftern world, the legend runs that the eldeft daughters of the Pileck family, in Poland, are transformed into doves if they die unmarried, into owls if married, at their death. The ftudent of language and myths will find much food for thought in thefe notices of Shakefpeare's "clamorous owl." There is a Flemifh painter, Henri de Bles, born 1480, who always painted an owl in his pictures, and was thus called "Civetta." A picture bought for the National Gallery in 1882, from the Hamilton collection, was faid to be by this painter, but clofer infpection fhowed that the fo-called owl was a vulture.

Until the rife of a fchool of nature-loving poets, beginning with Gilbert White at the end of the eighteenth century, the owl was only treated by the poets as a bird of night and terror. It was a fynonym for all that is moft ill-boding and fearfome. In the fo-called Chaucer's "Romaunt of the Rofe," the owl "of deth the bode ybringeth."

[1] See the *Saturday Review*, Feb. 4, 1882, on Rolland's "Faune Populaire de la France," and Kelly's "Indo-European Folk Lore," p. 75.

In Shakespeare it is "the baker's daughter," by a
seeming confusion of folk-lore with the woodpecker. It is "the fatal bellman, which gives the
stern'st good-night;" the "boding scritch owl;"
the "ominous and fearful owl of death." When
it appeared by day (as the barn-owl often does),
its character only seemed the blacker:

> " The bird of night did sit
> Even at noonday upon the market-place,
> Hooting and shrieking."—(*Jul. Cæs.*, i. 3.)

King Lear, when he would fly from men to
dwell among the direst and most cruel of
creatures, determines "to be a comrade with the
wolf and owl" (ii. 4). Spenser also places the
poor owl in ill company, when, as Guyon and
the Palmer sailed together:

> "Suddeinly an innumerable flight
> Of harmefull fowles about them fluttering cride,
> And with their wicked wings them ofte did smight,
> And sore annoyed, groping in that griesly night.

> " Even all the nations of unfortunate
> And fatall birds about them flocked were,
> Such as by nature men abhorre and hate ;
> The ill-faste owle, death's dreadfull messengere ;
> The hoars night-raven, trump of doleful drere ;
> The lether-winged batt, dayes enimy ;
> The ruefull strich, still waiting on the bere ;
> The whistler shrill, that whoso heares doth dy ;
> The hellish harpyes, prophets of sad destiny ;

> " All those, and all that els does horror breed,
> About them flew, and fil'd their sayles with feare." [1]

Another and a different view of the bird is taken
by Daniel. This scene of the little birds flouting

---

[1] " Faerie Queene," ii. xii. 35.

the owl, is one that muſt have been noticed by
moſt lovers of the country:

> "Look how the day-hater, Minerva's bird,
>   Whilſt privileged with darkneſs and the night,
>   Doth live ſecure t' himſelf, of others feared.
>   If but by chance diſcovered in the light
>   How doth each little fowl (with envy ſtirred),
>   Call him to juſtice, urge him with deſpite,
>   Summon the feathered flocks of all the wood,
>   To come to ſcorn the tyrant of their blood !"[1]

Owls had a diſtinct medicinal value with the
Romans, as indeed had almoſt every bird and
plant known to them. In this connection folk-
lore is ſeen allying itſelf with ſcience, as yet crude
and fanciful. "The feet of a ſchriche Owle burnt
together with the herb Plumbago, is very good
againſt ſerpents. But before I write further of
this bird," adds Pliny, "I cannot ouerpaſſe the
vanitie of Magicians which herein appeareth moſt
euidently; for ouer and beſides many other
monſtrous lies which they haue deuiſed, they giue
it out, that if one doe lay the heart of a Scrich-
Owle on the left pap of a woman as ſhe lies
aſleep, ſhe will diſcloſe and utter all the ſecrets of
her heart; alſo, whoſoeuer carie about them the ſame
heart when they go to fight, ſhal be more hardie,
and performe their deuoir the better againſt their
enemies." Owl's eggs were reported to cure all
defects and accidents to which the hair was liable;
but, aſks Pliny indignantly, "I would faine know
of them what man euer found a Scrich-Owle's

---

[1] "Hiſtory of the Civil Wars," 99.

neſt and met with any of their egges, conſidering that it is holden for an vncouth and ſtrange prodigie to haue ſeen the bird itſelf?   And what might be he that tried ſuch concluſions and experiments, eſpecially in the haire of his head?"[1]

[1] Pliny, "Nat. Hiſt." (Holland), xxix. 4.

CHAPTER VI.

PYGMIES.

"Do you any ambaſſage to the Pigmies?"
("Much Ado," ii. 1.

IN the myths of antiquity, and in
modern folk-lore, pygmies hold an
equally honoured place. Thoſe of
early Greek legend are own brothers
to the trolls and elves of Northern mythology;
while their deſcendants, the pixies of to-day, yet
dance among the moonlit glades of Devon
and Cornwall in the belief of the Weſtern
peaſantry. Pygmies firſt appear in the "Iliad,"
iii. 2-7: "The Trojans advanced with clangour
and a war-cry, like birds; like the clamour of
cranes aloft in heaven, when flying from winter
and a mighty ſtorm, loudly clamouring, they wing
their way to the ocean ſtreams, bringing ſlaughter
and death to the Pygmies." Ariſtotle ("Hiſt.
An.," viii. 14, 2) amplifies this paſſage, which
he evidently had in his remembrance: "Cranes

migrate from the Scythian regions into the marfhes of Upper Egypt where the Nile takes its rife. The Pygmies dwell in thefe parts ; the tale told of them is no myth, but in good truth they are a nation of fmall ftature, as the ftory runs, both they and their horfes. They live after the fafhion of Troglodytes." Strabo naturally diffents from the common belief: "The Ethiopians lead a wretched life, and are for the moft part naked, and roamers from place to place. Their flocks confift of fmall fheep, goats, oxen, and dogs. They are morofe, too, and warlike, in confequence of their fmall ftature. Perhaps it was from the fhort ftature of thefe men that the ftory of the Pygmies was devifed and ftruck out. No one worthy of credit relates that he had actually feen them."

The ordinary ftory appears in Pliny that they fit upon rams and fhe-goats, and, armed with arrows, in the fpring-time defcend in a body to the fea, and eat the eggs and young ones of the cranes, an expedition which occupies three whole months. He places the Pygmies among the furtheft nations of India. With him agrees Ctefias, who ftates that in the centre of India are men of a dark hue called Pygmies, ufing the fame language as the reft of the Indians. They are covered with long hair, and very fmall, the talleft being two cubits in height, but moft of them only one and a half cubit in ftature. Such ftories probably helped Swift to his Lilliputians, who alfo bore bows and quivers full of arrows.

G

" The nation of the prettie Pygmies," adds Pliny,[1]
" enjoy a truce and ceffation from armes every
yeare, when the cranes, who ufe to wage war
with them, be once departed and come into our
countries." Vefpafian, at the dedication of the
Coloffeum, prefented a fpectacle to the people of
a battle between cranes and a number of dwarfs
who imitated Pygmies.

Aulus Gellius gives a fimilar account of Pygmies,
placing them in India, and making the talleft of
them but two feet and a quarter in height.[2]
Hanno, in his " Periplus," places them in the Atlas
mountain, and ftates that they " run fafter than
horfes," and are Troglodytes. Meffrs. Hooker
and Ball, during their recent travels in the Great
Atlas, obferved several Troglodytic habitations.
Juvenal amufingly comprehends all the learning
of the ancient world on Pygmies in " Sat." xiii.
167-170, and of their army fays, " Tota cohors
pede non eft altior uno " (173).

Sir Thomas Browne has no difficulty in his
"Vulgar Errors " (Book iv. 11) in difpofing of
thefe fables after his own fafhion. Having mentioned the above paffages, and feveral others from
ancient poets and writers, he concludes that what
was "only a pleafant figment in the fountain,
became a folemn ftory in the ftream, and current

[1] " Nat. Hift.," x. 24 (Holland).
[2] ix. 4, 10. Pliny afferts that the Pygmies live among the
marfhes where the Nile rifes, curioufly anticipating modern
geographical refearch. The Troglodytes, he places on the
Arabian Gulf next the Ichthyophagi, " of wonderful fwiftnefs,
fwimming like fifh " (vi. 30, 34).

ftill among us." Moft of his fcorn is poured out
upon Ariftotle, who can afford to fmile at it
however, "wherein indeed Ariftotle plays the
Ariftotle, that is, the wary and evading affertor ;
for though with *non eft fabula*, he feems at firft to
confirm it, yet at the laft he claps in *ut aiunt*,
and fhakes the belief he put before upon it."
Much of his own chapter is taken up with a
confideration of Ezekiel xxvii. 12, where, in the
Vulgate, the Pygmies appear as a tranflation of
"Gammadim," which our verfion tranflates "men
of Arvad :" "Et Pygmæi qui erant in turribus
tuis pharetras fuas fufpenderunt in muris tuis per
gyrum." It is difficult, indeed, to connect the
Pygmies with the city of Tyre, to which thefe
words refer ; fome might call it impoffible, were
not the commentary of the ingenious Forerius
extant. He confiders that "the watchmen of
Tyre might well be called Pygmies, the towers of
that city being fo high that, unto men below, they
appeared in a cubital ftature." But the Pygmies,
it will be feen, are to be found in much ftranger
places than ancient Phœnicia ; fuffice it now to
ftate Sir T. Browne's cautious judgments on them :
"Since it is not defined in what dimenfions the
foul may exercife her faculties, we fhall not conclude impoffibility ; or that there might not be a
race of Pygmies, as there is fometimes of giants ;
but to believe they fhould be in the ftature of a
foot or a fpan requires the preafpection of fuch a
one as Philetas the poet in Athenæus, who was
fain to faften lead unto his feet, left the wind

fhould blow him away." Of courfe Milton, with
his claffical lore, has not forgotten

> " That fmall infantry
> Warr'd on by cranes."—(*Par. Loft*, i. 575.)

For a later difquifition on Pygmies, the reader
may be referred to Ritfon's differtation, publifhed
at the end of his "Fairy Tales" (London, 1831).
He, too, quotes the chief claffical allufions to
them, adding (from Ctefias): "Of thefe Pygmies
the King of the Indians has 3,000 in his train,
for they are very fkilful archers. They are, however, moft juft, and ufe the fame laws as the other
Indians." Sir John Maundeville plants them near
the "gret ryvere that men clepen Dalay;" calls
them three fpans high, "thei lyven not but fix
yeer or feven at the mofte, and he that lyvethe
eight yeer, men holden him there righte paffynge
old. Thefe men be the worcheres of gold, fylver,
cotoun, fylk, and of alle fuch thinges of ony other
that be in the world," with more marvels.[1] One
Mr. Grofe, fays Ritfon, author of "A Voyage to
the Eaft Indies" (London, 1772), had heard of
Pygmies in Coromandel, but foon after, to his
amazement, he difcovered them in Great Britain.
"At the north poynt of Lewis there is a little ile
called the Pygmies ile, with ane little kirk in it
of their own handey-wark, within this kirk the
ancients of that countrey of the Lewis fays, that
the faid Pigmies has been eirdit thair. Maney
men of divers countreys has delvit upe dieplie the

---

[1] "Voiage and Travaill" (London, 1727), p. 232.

flure of the litle kerke, and i myſelve amanges the leaue, and has found in it, deepe under the erthe, certain banes and round heads of wonderful little quantity, allegit to be the banes of the ſaid Pigmies, quhilk may be lykely, according to ſundry hiſtorys that we reid of the Pigmies; but i leave this far of it to the ancients of Lewis."[1]

In the *Academy* (March 19, 1881) may be found an account of three modern Pygmies from Africa, the only repreſentatives of their race now living in Europe. The two boys are at preſent being brought up under the protection of Count Miniſcalchi at Verona, while the girl is leſs fortunately placed at Trieſte. The elder boy, Thibaut, now meaſures 1·42 mètre (55·9 inches) in height, and is believed to have reached his greateſt ſtature. He is probably about nineteen years of age. Chairallah, on the other hand, is ſtill growing, and at preſent meaſures 1·41 mètre (55·5 inches); he is ſuppoſed to be about fifteen years of age. Theſe lads have very pronounced dolicocephalic ſkulls, with the characteriſtic three-lobed form of noſe. Their prognathiſm is very ſtriking; the mouth large; the lips thick; the teeth ſtout, well-ſeparated, and exceedingly white. Tufts of black woolly hair have appeared upon the cheeks, the chin, and the upper lip of Thibaut. Chairallah, on the contrary, ſhows no trace of hair upon the face; his viſage, however, has greatly lengthened with age. They have forgotten both

---

[1] "Deſcription of the Weſtern Iſles of Scotland." By Donald Monro (ed. 1784), p. 37. Martin ſays of this that the natives call theſe the bones of Luſtbirdan (*i.e.* pygmies).

their native Akka and the Arabic which they learnt when young, but fpeak, read, and write Italian. The girl can neither read nor write, but can fpeak Italian, and a little German, languages which fhe hears daily around her. She is fuppofed to be about fifteen years of age. Her prefent height is 1·34 mètre (52·7 inches[1]). All thefe three Akkas have good health, and are generally well-behaved, but have exceedingly childifh taftes.

If thefe different accounts of ancient and modern Pygmies be weighed, it will be found that either monkeys or aboriginal Troglodytic tribes are defcribed under that name. The Book of Job[2] alludes to the Horites and other Troglodytic races of Paleftine, whofe haunts were in the rocks of Edom on one fide of the world. On the other fide, the neolithic Iberian race, the Trog-lodytes of Dordogne, the Piðs, makers of the fo-called Piðs' houfes, the primal natives of the Atlas, and the like, have been efteemed Pygmies by the races which fucceeded them. In India the dark fkins and flat Mongol features of many of the aboriginal hill and jungle tribes being diftafte-ful to their Aryan conquerors, led the latter to transform them into goblins, pygmies, or demons. In juft the fame manner the aborigines of Scan-dinavia became the elves and gnomes, the mif-chievous trolls or pygmies of Icelandic and Norfe tradition. Doubtless the fcorn of the conquerors

---

[1] Cnf. Juvenal, "Satires," vi. 504. "Breviorque videtur Virgine Pygmæa."
[2] Cap. xxx. 6, 7.

as well as their proud fupremacy were thus flattered. In depreciating their forerunners, they exalted themfelves. The procefs by which thefe Indian and Oriental fables paffed into Europe, and what in fome cafes is ftill more important, the Buddhift origin of thefe Oriental fables themfelves, has been pointed out by the late Theodor Benfey. It is curious that the ideas of claffic poets, on the degeneracy of the human race, are being every day contradicted by the difcoveries of fcience. Not leaft among thefe corrections of popular beliefs is the evidence for the gradual amelioration of mankind to which the legends and hiftory of fo-called Pygmy tribes teftify. They corroborate the teftimony of revelation and the infight of modern poets, more true in this particular than their brethren in the paft, that there is a golden future for the race, an "increafing purpofe running through the ages." Material progrefs, in fhort, means in moft cafes the moral and mental advancement of man. Civilization is a light whofe radiance is ever piercing further and further into the realms of darkness:

> "Wait ; my faith is large in Time,
> And that which fhapes it to fome perfect end."

As for any other theory, "a pygmy's ftraw doth pierce it" ("King Lear," iv. 6).

CHAPTER VII.

ELEPHANTS.

"Th' unwieldy elephant,
To make them mirth, us'd all his might, and wreath'd
His lithe probofcis."—(*P. L.* iv. 345.)

THE elephant family runs far back into
the pliocene age. It attained a large
range in poft-pliocene times. At
prefent, it is well known, we poffefs
two main branches of the ftem in the Indian and
African elephants, which are well marked off from
each other. The latter kind is not now tamed,
but it is fuppofed that the elephants ufed by the
Carthaginians were of this fpecies. Singularly
enough, the extinct probofcideans alfo fall under
two divifions, the elephant proper, of which the
mammoth (*elephas primigenius*) is the type; and
the maftodon, diftinguifhed by its udder-like teeth,
adapted for bruifing coarfer vegetable fubftances,
and the prefence of two tufks in the lower jaw
of both fexes. Species of the maftodon lived in
Europe, Auftralia, and America. Owen's *M.
Anguftidens* has been found in our English Crag,

and the bones of *M. giganteus*, from the United States, may be feen ſet up at the Britiſh Muſeum. This creature is 20 feet 2 inches long, and 9 feet 6¾ inches high. Dr. Falconer diſcovered the remains of feveral ſpecies alſo in the Sewalik Hills, Eaſt India. There, too, he found bones belonging to ſix ſpecies of true elephant. Reverting to the mammoth, there is no need to dwell upon the diſcovery of the carcafe of one almoſt wholly preſerved, fleſh, ſkin, and hair, in 1799 on the banks of Lake Onkoul in Siberia. The mammoth was very widely diſtributed over the globe, being a pre-glacial as well as a poſt-glacial inhabitant of Great Britain. Numerous remains of it have been found in Norfolk and Suffolk, and dredged out of the German Ocean. It was the lateſt form of the extinct elephants. Its tuſks have for ages in their foſſil ſtate ſupplied almoſt all the ivory that Ruſſia uſes. Another inſtance occurred in 1846 of a frozen mammoth being thawed out of the ice, but it could not be preſerved. Lieutenant Benkendorf and a party of Coſſacks, in that year diſcovered it on the banks of the Indighirka, ſtanding upright in the ice-bound tundra in the place where it had been bogged, and narrowly eſcaped being themſelves ſwept out into the Arctic Sea, as the thaw proceeded, together with the mammoth, which actually was ſo loſt. Several points in the hiſtory of the mammoth ſhow that it probably was the anceſtor, geologically ſpeaking, of the preſent Indian elephant. The oldeſt carving known, found in the Madelaine Cave, in the

Dordogne, by M. Lartet, reprefents the mammoth on a piece of its own ivory.[1]

The royal pupil of Ariftotle put him in pofleffion of a good deal of knowledge on the elephant. The philofopher fpeaks of it, from fifteen which were captured at Arbela, as the tameft and mildeft-tempered of creatures, full of intelligence, and living to the age of 120 or 200 years ; but at its beft when 60 years old. He knew its abhorrence of cold, too. Preconceived notions, however, come in when he ftates that it paffes through rivers, wading in them as far as the end of its trunk allowed, for it breathes through this, and cannot fwim on account of the bulk of its body. On the contrary, the elephant is a capital fwimmer, and delights in nothing fo much as deep water. The Mahouts frequently caufe their charges at the prefent day to fwim over wide rivers, and even the Ganges. He fpeaks alfo of the time of *muft* in male elephants ; how, at thofe periods, they are in a ftate of madnefs, and knock over houfes as if they were badly conftructed, and commit all manner of exceffes. " They tell that fcantinefs of food renders them tamer, and by bringing up to them other elephants they reftrain them by ordering thefe to beat them." He fpeaks, too, of olive oil being given elephants to expel any piece of iron they may have accidentally eaten ; and has a chapter on their ailments. The food of an elephant is meafured by him almoft with the exactnefs with which the keep of Jumbo was cal- .

[1] Wilfon, "Prehiftoric Man," i. p. 107.

culated at £500 per annum, when that much-puffed beaſt was ſold to Barnum. It can eat, he informs us, nine Macedonian buſhels at a meal, and ere now had been known to drink fourteen Macedonian meaſures at once. He notes, alſo, that ſome are fiercer and more courageous than others, a faƈt well known at preſent to all Indian tiger-hunters, and that they puſh over palm-trees with their foreheads, then walk up them and eat what they deſire of them. This, too, is a habit confirmed by all modern travellers. His account of elephant-catching compriſes in one ſhort paragraph the whole of the beſt modern books on the ſubjeƈt, whether Tennant or Sanderſon. "The chaſe of elephants is on this faſhion. Men mount ſome of the tame and courageous elephants and purſue the herd. When they have come up with it, they bid their own animals to beat the wild ones with their trunks, until they give in through faintneſs. Then the elephant-taker leaps on one of them and guides it with his weapon. After this it ſoon becomes mild and ſubmiſſive. When the elephant-taker has mounted them, all are in ſubjeƈtion; but when he has diſmounted ſome remain ſo, while others return to a wild ſtate. While theſe are raging, they fetter with chains their front legs, in order that they may be quiet. Both ſmall and great are thus captured."[1]

The firſt twelve chapters of Pliny's eighth book of "Natural Hiſtory" contain almoſt all the faƈts as well as the fancies known to the ancients about

[1] Ar., "Hiſt. Anim.," ix. 33 ; vi. 17 ; viii. 25, 11 ; ix. 2.

elephants, folk-lore and fcience not yet being feparated in the cafe of natural productions. Certain tribes of Africa fubfifted by hunting elephants, he tells us, and the city Ptolemais was built by Philadelphus for the fake of enabling him to hunt elephants.[1] Certain of thefe African elephants are faid to affemble by fours and fives in the maritime diftricts of Ethiopia, and having interlaced their legs and trunks, with erect heads and ears, to commit themfelves to the waves, by which they are floated over to the finer pafturage of Arabia. As for the Indian elephants, life is made a burden to them by the huge ferpents which wrap their coils round them. The elephants, however, undo thefe coils by their trunks, whereupon the ferpents faften round them by the tail, and thrufting their heads into the elephant's noftrils, ftop their breath, and fting them internally to death. Another account tells how the ferpents lie in wait in the water where elephants come to drink, and then, feizing their trunks, fting them in the ear, the only part which they cannot defend by their trunks. The Troglodytes, too, lie in wait up trees till the laft of the herd is paffing underneath. Upon this they drop, and feizing its tail with the left hand, hamftring it with a fharp weapon in their right. A fimilar mode of ftealing on elephants and hamftringing them is ftill purfued in the Eaft. Their battles with Pyrrhus fhowed the Romans that an elephant's trunk could eafily be cut off; and the

[1] "Nat. Hift.," vi. 34.

celebrated combat of a Roman captive with a Carthaginian elephant in the arena before Hannibal on the latter's promife that the flave's life fhould be fpared did he prove victor, conclufively proved that thefe creatures need not be greatly dreaded in war. The Romans were never fair to Hannibal, and Pliny cannot refrain from adding that Hannibal was fo chagrined at this difcovery that he sent horfemen, when the man had departed, to waylay and flay him. For accounts of Pompey and Cæfar exhibiting fhews of elephant-combats in the circus, and a multitude of curious particulars, we muft be content to refer the reader to the above citation, only warning him that many of the ftories told by Pliny require qualifying with the warning of the fhowman in a modern menagerie, who pointed out the porcupine, and obferved: "Buffon fays that he fhoots his quills; Buffon's a liar!" Ælian, Strabo, and Arrian are full of de-tails on elephant-catching and taming. Cicero was prefent at a *venatio* given by Pompey, B.C. 55, when twenty elephants were exhibited in the circus, and killed by darts. He adds that "a great admiration for the huge beafts fell on the fpectators, and no delight was taken in their death. Moreover, a certain feeling of pity followed the fpectacle, the populace not being able to withftand the opinion that there was a kinfhip to man in the fagacious creature."[1]

Elephants were firft feen in Italy at the invafion

[1] See Lord Cockburn on "The Chafe," *Nineteenth Century*, Dec. 1880 ; and cnf. Juvenal, "Satires," xii. 101-114.

of Pyrrhus, B.C. 280. He defeated the Romans
at the Siris in that year by their aid. Indeed, the
Roman army was only faved from annihilation in
that combat by one of the elephants, whofe trunk
had been cut off by a Roman foldier, turning
back upon and throwing his own party into con-
fufion. Ere long the Romans learnt to ufe them
in war, while exhibitions of their fagacity in time
of peace frequently amufed the populace at Rome.
It is curious that the elephant is never reprefented
among the Egyptian hieroglyphics, although it
was perhaps an inhabitant of Upper Egypt in
early times, where the ifland Elephantine remained
as an evidence of the fact.[1] Rawlinfon fuppofes
that elephants were firft ufed in the hiftory of
military fcience at the battle of Arbela,[2] " to which
they added an unwonted element of grotefquenefs
and favagery." They do not feem to have been
of much fervice in the actual ftruggle. Macaulay
has remembered the elephants of Pyrrhus in his
" Prophecy of Capys " :

> " The Greek fhall come againft thee,
>   The conqueror of the Eaft.
> Befide him ftalks to battle
>   The huge earth-fhaking beaft,
> The beaft on whom the caftle
>   With all its guards doth ftand,
> The beaft who hath between his eyes
>   The ferpent for a hand."

---

[1] See Rawlinfon, "Ancient Monarchies," iii., p. 148. In
the fourth century before our era the elephant withdrew to
India ("Chabas," p. 576).

[2] *Ibid.*, p. 387.

Thefe latter lines allude to the Lucretian epithet for the elephant, *anguimanus.*[1]

The recent excitement about the elephant Jumbo, which the Zoological Society fold to Barnum, called forth many interefting notices of the elephant in the Middle Ages in the public prints. From thefe the following facts may be culled. In the year 1229 an elephant was fent by the Soldan of Babylon as a rare prefent to the Emperor Frederick II. But it was not until 1255 that the firft elephant was feen for the lapfe of 1200 years in Britain. This was prefented by the King of France, as we learn from the chronicles of John of Oxenedes and others. It was houfed in the Tower, and lived on till the forty-firft year of Henry III., A.D. 1257, when it feems to have died, aged only twelve years. The charges of itfelf and keeper may be feen in the " Chancellor's Roll." Juft as Jumbo has immortalized his keeper, Scott, fo this royal elephant-keeper ftill lives as John Gouch. The fheriff of Kent was commanded together with him " to provide for bringing the King's elephant from Whitfand to Dover."[2]

It feems that white elephants have an actual exiftence apart from proverbs and effays. Mr.

---

[1] Compare Lucretius, ii. 538 :

> " Elephantos India quorum
> Millibus e multeis vallo munitur eburno
> Ut penitus nequeat penetrari."

They were known alfo to Lucretius as " boves Lucas " (v. 1301).

[2] *Notes and Queries,* 6th S., v. 385.

Bock, a recent traveller, was in 1880 very kindly received by the King of Siam, and witneſſed the proceſſion of the ſacred white elephants. The ſkin of theſe ſo-called white elephants he deſcribes as being rather a pinkiſh-grey. He made a coloured drawing of the lateſt addition to the royal ſtables, with which the King was much pleaſed.

A much-lauded white elephant arrived in London in Jan., 1884, but greatly diſappointed moſt people. It was light-coloured, and ſpotted on the root of the trunk and over the ears. One authority looked upon theſe markings as being the reſult of albiniſm, another as being due to a diſeaſe known as *leucoderma* (Prof. Flowers and Mr. B. Squires's letters to the *Times* of that date). Conſiderable ſanctity is attached to white elephants in the Eaſt. The Hindoos perhaps connect them with Airawata, the elephant of India, from whom the great river Irawadi, or Iravati, derives its name, like the Hydraotes, or Ravee, of the Punjaub. After a ſhort time this ſo-called white elephant left us for the Americans, a people more appreciative of ſuch marvels.

The elephant does not appear in the Homeric poems, but ivory is often mentioned. A celebrated paſſage ("Iliad," iv. 141) compares the blood on Menelaus, when wounded by an arrow, to a Mæonian or Carian woman ſtaining ivory with crimſon to be an ornament for horſes' heads, "and it lies in her chamber, and many horſemen deſire to wear it, but it ſtays as an ornament for a

king." [1]　The Trojan reins were ornamented with
ivory ſtuds ("Iliad," v. 583).　In the "Odyſſey"
more uſe is made of it.　Athene makes Penelope
"whiter than new-ſawn ivory" (xviii. 196).　Her
chair was deftly wrought with ivory and ſilver
(xix. 56), and the key of her chamber had an
ivory handle (xxi. 7).　We hear, too, of a bronze
ſword with ſilver handle and ſheath of freſh-ſawn
ivory (viii. 404); while in the palace of Menelaus
at Sparta were bronze, gold, amber and ivory, like
the halls of Olympus (iv. 73).　Perhaps the moſt
celebrated alluſion to this ſubſtance, however,
occurs in Penelope's account of the twin gate of
dreams, "the one of which has been faſhioned of
horn, and the other of ivory.　The dreams which
paſs through the ſawn ivory are deceptive, bringing words which have no fulfilment, but thoſe
which proceed through the poliſhed horn bring
true iſſues, whenever a mortal ſees them" (xix.
564). [2]　Virgil often touches on ivory in much the
ſame connection as the above.　"India mittit
ebur," he tells us ("Georg.," i. 57); and at Cæſar's
death, among the dread portents, "Mæſtum illa-
chrymat templis ebur" ("Georg.," i. 480).　He
ſpeaks, too, of "dona auro gravia ſectoque
elephanto" ("Æneid," iii. 464).　The Aſſyrians
carried on a great traffic in ivory with the Eaſt,

---

[1] Compare Virgil, "Æneid," xii. 68 :
　　"Indum ſanguineo veluti violaverit oſtro
　　Si quis ebur."
[2] Homer here indulges in a play on words.　The ivory is
ἰλίφας; the word for "deceptive" is ἰλεφάιροντα.　Virgil imitates the whole paſſage at the end of the ſixth "Æneid," (893).

H

and diffeminated it through the Weftern countries and Europe. The native country of the Eaftern elephant is the peninfula of India. Egyptian ivory was largely brought from Ethiopia, though their elephants were originally from Afia.[1]

Sir Thos. Browne has a fenfible chapter in the main on elephants, in his "Vulgar Errors," condemning feveral "old and gray-headed errors" on it. His own credulity, however, is amufing to the prefent generation, efpecially when he deems it ftrange that the curiofity of man, which had tried to induce many beafts to fpeak, had never attempted to tutor an elephant, for "the ferpent that fpake unto Eve, the dogs and cats that ufually fpeak unto witches, might afford fome encouragement."

The elephant occafionally appears upon coins ; as on one of Tarentum, probably connected with the invafion of Pyrrhus ; alfo on one of Vefpafian. It is found, too, on the coins of Metellus, who brought many Carthaginian elephants to Rome in the Firft Punic War ; alfo on thofe of Cæfar, from the legend that that name was the Carthaginian word for an elephant, and was originally applied to the firft of the Julian gens who had flain one of thefe creatures. It meant, as a fymbol on a coin, eternity ; and fometimes munificence in giving games to the populace. Cæfar is amufingly connected, by the Rev. J. Coleridge, a man of

---

[1] See "Dict. of Bible," *sub voc.* "ivory." Polydore Vergil has a proverb alluding to the flow geftation of elephants— "citius elephanti parient."

fome learning for his day, and father of the poet,
with elephants. It feems that, about the middle
of laft century, much curiofity was fhown with
regard to the foffil elephant bones and ivory fo
often found in South Eaftern England, and there
were many fpeculations about the manner in
which elephants could have reached our fhores.
In July, 1757, that clergyman (who was Vicar of
Ottery St. Mary, Devon) wrote his views to the
"Gentleman's Magazine." A previous corre-
fpondent had hazarded the notion that the
Romans had brought over thefe huge creatures to
intimidate the Britons; but, he adds, "we have
not the leaft account of any fuch thing." Mr.
Coleridge, however, points out that a paffage in
the "Stratagems" of Polyænus expreffly mentions
that an elephant was brought over by Cæfar and
ufed in forcing the paffage of the Thames when
the Romans were oppofed by Caffivelaunus. The
Romans then caufed their elephant to advance,
wearing an iron coat of mail, and carrying bow-
men and flingers in a little caftle on its back,
whereupon the Britons at once fled. Cæfar, he
adds, probably omitted this account in his
"Commentaries," thinking that the mention of it
would detract from the honour of his victories.
But the clofing fentences of the letter are fo interefting from the ftandpoint of the geologift in
the nineteenth century, that it is worth while
quoting them: "It is reafonable to fuppofe that
as the Romans reaped fuch advantages from one
elephant, they would bring over more of thefe

animals with them, and, as the Roman conquefts
were chiefly about Suffex, Effex, and Kent, it is
moft likely that the bones of thefe creatures fhould
be found in thofe counties. It cannot be proved,
indeed, that thefe bones have not lain ever fince
the general flood; but an hiftorical truth is, in my
opinion, preferable to any hypothefis whatfoever."
Modern fcience can well afford a fmile at the
amufing candour of thefe conclufions.

In the Eaft, as is only natural, the elephant
being regarded as poffeffed of more than mere
brute wifdom, is often deemed facred. Thus the
Hindoo Ganefha (god of wifdom) is reprefented
with an elephant's head, and the creature itfelf
frequently appears in the art of Hindoftan. It
is very rarely feen in Englifh architecture; but
an elephant's head and trunk are fculptured on
one of the pillars of the North or Dorfet Chapel
of the Church of Ottery St. Mary, Devon. On
the fummit, too, of Gofberton Church, Lincolnfhire, appears an elephant with a huge fpiral trunk.
In the fo-called Pictifh ornamentation on ancient
Scottifh fculptured ftones, a good many obfervers
have fancied that they could detect the elephant's
form, and efpecially the fpiral of its trunk.
Doubtlefs much of this is due to imagination.
In fome cafes there may be a faint remembrance
of the mammoth. Elaborate fchemes of mythical
orientalizing have been founded on this fpiral line,
which, after all, is fimply a characteriftic mark
of early Scottifh ornamentation. The late Dr.
Burton fays, " It is pretty evident, when we in-

ſpeċt him cloſely, that the animal ſo often ſuppoſed
to be figured on ancient Scottiſh ſculptured ſtones,
though a ſtrange beaſt of ſome peculiar conventional
type, is no elephant. That ſpiral winding-up of his
ſnout, which paſſed for a trunk, is a charaċteriſtic
refuge of embryo art, repeated upon other parts
of the animal. It is neceſſitated by the difficulty
which a primitive artiſt feels in bringing out the
form of an extremity, whatever it may be—ſnout,
horn, or hoof. He finds that the eaſieſt termina-
tion he can make is a whirl, and he makes it
accordingly. Thus the noſes, the tails, the feet
of the charaċteriſtic monſter of the ſculptured
ſtones all end in a whirl. The ſame difficulty is
met in repeated inſtances in theſe ſtones by another
ingenious reſource. Animals are united or twined
together by noſes or tails, to enable the artiſt to
eſcape the difficulty of executing the extremities of
each ſeparately."[1] Theſe remarks are perhaps more
ingenious than convincing when we remember
the extreme love for the ſpiral and for convoluted
and parallel ornamentation which extended into
the Saxon and Norman decoration of churches.
There was doubtleſs a myſtic ſignification attached
to the many curious ſpiral lines of early North-
Britiſh ſculptures.

Much information has recently appeared reſpeċt-
ing the mammoth, which will here be condenſed.
The Arabs in the ninth and two ſucceeding centuries ſhowed immenſe enterpriſe and energy, their

[1] "The Bookhunter," p. 399; and ſee "The Ancient
Sculptured Stones of Scotland" (Spalding Society, 2 vols. fol.).

traders frequenting the borderlands of Siberia, and probably firft initiating the trade in foffil ivory throughout the Weft. There is every probability that the very name " mammoth," as well as "mammoth ivory" itfelf, were firft brought to the Weftern world by the Arabs. " Mammoth" is merely a form of "behemoth." Witzen, who firft defcribed the creature in 1694, ufes the two names as fynonymous; and Father Avril, a Jefuit who travelled in China in 1685, calls the mammoth " Behemot." The Turkifh dialeḉts habitually interchange *b* and *m*, and there feems no doubt that Job's " behemoth," which the Arabs pronounce " mehemot," filtered through the Ruffian and Tartar tongues into our " mammoth," the word " behemoth" being ufed of any monftrous beaft originally, and then confined in the North to the great foffil elephant.[1] The creature itfelf was firft defcribed by Witzen (whofe book, written in Dutch, has never been tranflated) in 1686. The firft mammoth tufk was brought to England by Jofias Logan in 1611, and had been purchafed near the Petfchora river. A mammoth mummy was firft difinterred about 1692; another was found near the river Alafej in 1787; next comes the one above defcribed in 1699 on the Tamut Peninfula; another was opened out on the Yenifej in 1839, and again others were found in 1846 and 1866.[2]

[1] See an excellent paper on the name " Mammoth" by H. H. Howorth, F.S.A., in *The Field Naturalift*, July, 1882, p. 30.

[2] "Voyage of the Vega," by Nordenfkiold (1881). See vol. i., p. 400 *feq.*

CHAPTER VIII.

THE HORSE.

"Ripa nutritus in illa,
Ad quam Gorgonei delapfa eft penna caballi.'
(Juv. *Sat.*, iii. 117.)

EMAINS of the horfe in a domefticated ftate have been found in Swifs
lake-dwellings of the Neolithic
period, but Profeffor Huxley deems
that the *anchitheres* of the upper Eocene times
were the true anceftors of the horfe. Thefe foffil
creatures were about the fize of Shetland ponies,
and poffeffed three diftinct hoofs on each foot.[1]
Without committing ourfelves to a belief in the
Darwinian doctrines of defcent, we may well be
grateful to fcience for pointing out the different
ftages in which creative Wifdom was pleafed
to fafhion fimilar extinct animals, before giving
man fo ufeful a creature as the horfe. A very
early fpecimen of art reprefents the foffil horfe

[1] Dawkins, "Early Man in Britain," p. 32, and Sir J.
Lubbock, "Addrefs to the Britifh Affociation," Sept. 1881 ;
fee, too, his " Fifty Years of Science," p. 9 (Macmillan, 1882).

carved on a rib by the cave-men of Dordogne, apparently with a flint graver.[1] The manner in which one horfe is reprefented as biting the tail of another at the fame time that it depreffes and puts back its own ears, is remarkably true to nature, and feems the fketch of an artift fkilled in the ufe of the pencil, rather than the fcratching of a favage. The *equidæ*, as a family, only date from Pliocene times. The foffil horfe of our iflands was the fize of a fmall horfe at prefent, and had a larger head than the domefticated races, as may be well feen in the engraving of the carved rib from Dordogne in Mr. Wilfon's book. Two or three fkeletons of horfes have been found in Scotland buried along with their owners, chiefs in the iron period, and the bridle-bits of thefe horfes are frequently very beautiful.[2] But with regard to horfe furniture, two moft fingular horfe-collars of ftone were found near the parallel roads of Glenroy in Scotland.[3] Thefe are models rather than the actual collars which were ufed in the ftone period, and are finely polifhed. Of courfe their difcovery led to much wild fpeculation about the parallel roads having once been the fcene of public games and chariot races, after the old-fafhioned type of archæology. Careful breeding has given the domefticated horfe both fize and fymmetry. We have feen Roman horfe-fhoes, found in Devon, which are very fmall compared with thofe ufed

[1] Wilfon, "Prehiftoric Man," i., p. 106 (1876).
[2] Wilfon, "Prehiftoric Annals of Scotland," 1851, p. 458 (feveral figures).
[3] *Ibid.*, p. 156.

for our preſent horſes; and Wilſon ſtates[1] that horſe-ſhoes found on the field of Bannockburn and at Niſbetmuir are remarkable for their very diminutive ſize. As for horſe-ſhoeing, the art was known in the time of Cæſar both to Britons and Saxons, although it is generally aſſerted to have been introduced into England by William the Norman. The Greeks were accuſtomed to nail a rim of iron on a horſe's hoof, as may be gathered from a Greek coin, now in the Britiſh Muſeum, found at Tarentum, and ſuppoſed to date from B.C. 200. The later Roman horſe-ſhoe, made of gold, which horſes wore and kicked off in triumphs, proceſſions and the like, were probably not nailed on the foot.

Paſſing from the animal's derivation to that of its name, as being a common domeſtic animal of the Indo-European races, it is not ſurpriſing to find the word "horſe" ſubſtantially one and the ſame in all the Aryan dialects. Thus it is aſva in Sanſcrit, ἵππος in Greek, and (connected by the dialectical ἵκκος) *equus* in Latin; "hors" (the Anglo-Saxon name), or "ors," by a uſual metatheſis became "ros" or "roſs" in German. The horſe was not uſed by the Jews until the times of David and Solomon, in conſequence of the hilly nature of their country, and becauſe of the direct prohibition (Deut. xvii. 16). It came to Paleſtine from Egypt, where it had been pro-bably introduced by the Hykſos. Thus it is not found repreſented on the monuments before the

---

[1] Wilſon, "Annals," p. 437.

eighteenth dynafty, and the agreement between its name in Egyptian and in Hebrew points to a Semitic origin.[1] With the Greeks it was facred to Pofeidon, and the well-known legend of his creation of the animal may either point to its introduction into Hellas by fea, or be an inftance of Greek poetic fancy (juft as we talk of "white horfes" when the waves ruffle the fea in fummer), and be connected with the horfes of the fun, fo frequent a myth in Oriental mythologies, which feem every morning to rife from the fea.[2] So the Rhodians ufed yearly to caft into the fea a four-horfe chariot which had been dedicated to the fun, and every ninth year in Illyricum four chariot horfes were fimilarly caft into the fea. Sophocles fpeaks of day dawning with its white horfes ("Ajax," 672).

Among the Perfians Mithra was the fun-god, and was perfonified, as alfo among the Greeks and Romans, as driving a team of horfes in his chariot. There are numberlefs allufions in ancient literature to the horfe as being an animal facred to the fun. "Perfia," fays Ovid, "appeafes the fun with a horfe that a flow victim may not be given to a fwift god."[3] Xenophon fpeaks of the fame facrifice. The Scythian Maffagetæ followed the fame cuftom, "facrificing the fwifteft of all mortal creatures to the fwifteft of the gods."[4] In

---

[1] Wilkinfon, "Ancient Egyptians" (Abridgment, vol. i., p. 386).

[2] To Neptune was attributed the invention of reins. Soph. "Œd. Col.," 713-15, Dind.

[3] "Faft.," i. 385.   [4] Lib. i. 216.

the Vedas the chariot of the ſun is drawn by two, ſeven, or ten horſes called "haritas," which is always a feminine noun. Profeſſor Müller has traced the connection between theſe and the Greek "charites" or "Graces," and the Greek god of love, Eros, with the Sanſcrit conception of Dawn.[1] The team of the ſun's chariot with the Greeks and Romans was four in number. No ancient ſculptor ever carved theſe prancing .fire-breathing ſteeds more nobly than has our own Gibſon in the wonderful *baſ-relief* to be ſeen at Wentworth Houſe, the divine youth reſtraining his plunging ſteeds without an effort, as it were, as the "wild team" ariſe

" And ſhake the darkneſs from their looſened manes,
And beat the twilight into flakes of fire." [2]

The Greek Hours who lead forth the chariot become in Sanſcrit oxen, from the notion of oxen going forth at morning to paſture, and returning with evening; and ſo, remarks Profeſſor Max Müller, we can underſtand the inner meaning of the old Homeric myth reſpecting the companions of Odyſſeus who killed the oxen of the ſun and never again ſaw their native land. They waſted their hours elſewhere, literally killed the time in idleneſs and voluptuous living. So, too, we can underſtand the force of the Homeric epithets applied to the ſun's horſes, "ſwift-flying," "ſwift-

[1] Max Müller, "Selected Eſſays," i., p. 439.
[2] Compare the horſes of the Sun in Virgil, " Æneid," xii. 113 :
" From the deep gulf the Sun's proud courſers riſe
And, rearing, from their noſtrils breathe forth flame."

footed," and the like, which Virgil follows in his "wing-footed" horſes.

It was probably due to ſome connection with the ſwift-flowing ſtreams of rivers that the ancients were often wont to ſacrifice horſes on their banks. Thus Xerxes, on croſſing the Strymon, when about to invade Greece with his enormous hoſt, ſacrificed white horſes to propitiate it.[1] And in much later times, while Vitellius offered the cuſtomary Roman ſacrifices by the Euphrates, its ſtream was appeaſed by Tiridates with the ſacrifice of a horſe.[2] Ten ſacred horſes of the celebrated Nyſæan breed were led, gorgeouſly capariſoned, before the chariot of Mithra on the march of Xerxes, while after it the royal chariot in which the King himſelf ſat in ſtate was alſo drawn by Nyſæan horſes. The Nyſæan plain, whence came the moſt prized horſes of the Perſians, was ſituated to the ſouth-weſt of Ecbatana, on the high uplands weſt of Mount Zagros. The Perſians have always been fond of horſes; indeed, their education, according to Herodotus, conſiſted in three things —learning to ride, to ſhoot, and to ſpeak the truth.[3] The pre-eminence of the Nyſæan horſes has now paſſed to the Arabian horſes of the Nedjd.

Ariſtotle ends a chapter about the age and dentition of the horſe, which might paſs muſter in a modern manual of farriery, with an account of a

[1] Herod., vii. 113.   [2] Tac. "Ann.," vi. 37.

[3] Herod., vii. 40 and i. 136; and Rawlinſon, "Five Great Monarchies," p. 145 and ii. p. 261.

celebrated superstition among the ancients, the *hippomanes*. "When a foal is born," he says, "the mother immediately bites off a growth upon its forehead, which is a little less than a fig in size, and is broad, circular, and black. If anyone is beforehand in obtaining this, and the mare should smell it, she is beside herself and maddened with its odour. Hence forceresses seek for and collect it as a charm."[1] And he adds, "The horse seems to be eminently an animal fond of its young; thus, when mares have lived together, if one dies the rest cherish its foal, and often the barren ones themselves cherish these foals, but by reason of having no milk kill them." Pliny evidently had Aristotle's book before him, but adds a multitude of fables, as his wont is, to the Stagyrite's common-sense. Thus Cæsar's horse would suffer no one but its master to mount it, and was notable for its forelegs ending in human feet. It was honoured with a gorgeous tomb, while at Agrigentum pyramids were erected as memorials of many horses. The great Semiramis was in love with a horse. The Scythian cavalry was famous; and on one occasion, when a chieftain was killed, his horse fell with tooth and hoof upon the victor and slew him. Such is the docility of the horse, that all the cavalry of Sybaris was taught to dance to the sound of a measure. It snuffs the battle afar off, and mourns its lost lord, sometimes even with tears. Nay, when King Nicomedes died his horse starved itself to death. When Dionysius left his

---

[1] Arist., "Hist. Animalium," vi. 22, 6 and ix. 5.

horfe foundered in a bog in order himfelf to efcape, the animal followed its mafter's footfteps with a fwarm of bees hanging on its mane; and in confequence of this portent Dionyfius feized upon the throne.   The fiercer the horfe, the deeper does he plunge his nofe into water when he drinks. Thefe and other ftill more wonderful myths, which are fcarcely to be told in the vulgar tongue, paffed current with the Roman encyclopædift for natural hiftory.[1]   He follows Ariftotle, too, in the marvellous ftory of the *hippomanes.*

Like moft of our domeftic animals, the horfe probably came into Europe from the vaft fteppes of Turkeftan and the Oxus.   Thence they formed the Spanifh ftock, which was fo celebrated amongft the Romans, and which Pliny commends for its well-ordered paces and high action.   The fame of Spanifh horfes, however, yet furvives ; and at the official entry of the Princefs Stephanie into Vienna on May 9, 1881, the day before her marriage, her carriage was drawn by milk-white fteeds of the pureft Spanifh blood.   Both the black and white varieties of the ftrain are fcrupuloufly kept pure in the Imperial ftud ; and, with the exception perhaps of the cream-coloured Hanoverians, are the only pure reprefentatives of the breed in exiftence.   The fwifteft African horfes alfo came of Spanifh blood. In Poland, buffaloes and wild horfes abounded in early times.   Full accounts of the Scythians on the fteppes of Southern Ruffia, and their nomadic mode of life with horfes and flocks, are given in Hero-

1 Pliny, " Hift. Nat.," viii. 42.

dotus. The Parthians, much farther to the Eaſt, were, if poſſible, ſtill more diſtinctively equeſtrian in their habits. "They are at all times carried on horſes. On them they fight, take their meals, perform all public and private duties, make their journeys, reſt, barter, converſe. The chief difference between ſlaves and freemen with them is, that ſlaves walk on foot, while freemen always ride."[1] A Roman poet, too, ſpeaks of "learning how many miles the Parthian horſeman can ride without water." Many mares of this ſtock were ſent into Macedonia by Philip, the father of Alexander the Great, to improve the native race. The cavalry of Thrace[2] and Theſſaly was famous with the ancients, and the mares, as in Arabia at preſent, were more highly valued than the horſes.

What the ancient ideal of a good horſe was may be gathered from Virgil :[3]

"Upright he walks, on paſterns firm and ſtraight,
His motions eaſy, prancing in his gait.
The firſt to lead the way, to tempt the flood,
To paſs the bridge unknown, nor fear the trembling wood.
Dauntleſs at empty noiſes, lofty-necked,
Sharp-headed, barrel-bellied, broadly-backed,
Brawny his cheſt and deep, his colour gray,
For beauty dappled, or the brighteſt bay,
Faint white and dun will ſcarce the rearing pay."

Yet the poſſeſſion of theſe points are of little avail without a long anceſtry; "let him trace

[1] See Juſtin, xli. 3; and Propertius, iv. 3, 35, quoted in Victor Hehn's "Kulturpflanzen und Hausthiere" (Berlin, 1877), p. 24.

[2] So Turnus,

"Maculis quem Thracius albis
Portat equus."—(" Æn.," ix. 49.)

[3] "Georg.," iii. 79, 121.

his breed to Epirus and warlike Mycenæ, and even deduce his pedigree from Neptune himfelf," then the refult is unmiftakable:

> "The fiery courfer, when he hears from far
> The fprightly trumpets and the fhouts of war,
> Pricks up his ears and trembling with delight,
> Shifts place, and paws, and hopes the promifed fight.
> On his right fhoulder his thick mane reclined,
> Ruffles at fpeed and dances in the wind.
> His horny hoofs are jetty black and round,
> His chine is double ; ftarting with a bound
> He turns the turf and fhakes the folid ground.
> Fire from his eyes, clouds from his noftrils flow,
> He bears his rider headlong on the foe."[1]

The line in which the Latin poet imitates the galloping of horfes, is well known to all lovers of the "Æneid." Another ftriking picture of the horfe when perifhing by an epidemic, merits quotation:

> "The victor horfe, forgetful of his food,
> The palm renounces and abhors the flood.
> He paws the ground and on his hanging ears
> A doubtful fweat in clammy drops appears ;
> Parched is his hide and rugged are his hairs.
> Such are the fymptoms of the young difeafe,
> But in time's procefs, when his pains increafe,
> He rolls his mournful eyes, he deeply groans
> With patient fobbing and with manly moans.
> He heaves for breath, which from his lungs fupplied
> And fetched from far diftends his labouring fide."

A drench of wine adminiftered through a horn has fometimes proved fuccefsful in arrefting the difeafe, but as often as not merely fupplied fuel for the flames:

> "For the too vigorous dofe too fiercely wrought
> And added fury to the ftrength it brought ;

---

[1] Dryden.

Recruited into rage he grinds his teeth
In his own flefh and feels approaching death.
Ye gods, to better fate good men difpofe,
And turn that impious error on our foes!"

Turning once more to the Eaft, we find the Affyrian horfes highly prized at prefent as they were of old. They are fmall of ftature, but of exquifite fymmetry and wonderful endurance. Mr. Layard mentions a cafe where a Sheikh refufed no lefs a fum than £1,200 for a favourite mare.[1] The Median horfes now belong to two diftinct breeds, the Turkoman, a large powerful animal with long legs and a big head, and the true Arabian, much fmaller and more perfectly fhaped. Of the Nyfæan horfes we have already fpoken. Babylonia bred vaft numbers of horfes under the Perfian rule. Thus one fatrap poffeffed 800 ftallions and 10,000 mares. The breed is thought to have been ftrong and large-limbed rather than handfome, the head being too large and the legs too fhort for fymmetry. The Huns, like the Parthians and Scythians, paffed all their lives on horfeback. Cilicia alfo poffeffed a breed of white horfes. It brought 360 of thefe—one a day for all the days of the Perfian year—year by year to Darius.[2] The horfes belonging to the lake-dwellers of the Pæonians were fed with fifh from the lakes below the pile-dwellings, according to Herodotus.[3] The Sigynnæ, a Thracian tribe in the extreme North, he alfo tells us, poffeffed horfes fo fmall that they

[1] Rawlinfon's "Five Empires,' i. 232 ; ii. 302 ; iii. 404; and Herod., i. 192.
[2] Herod., iii. 90.   [3] *Ibid.,* v. 16.

muſt have reſembled our Shetland ponies, with
hair as thick as five fingers. Theſe Lilliputian
animals were not ridden, but yoked to carts. It
is curious to find the father of hiſtory meaſuring
the depth of horſes' hair by fingers, when our
ſtandard meaſure for their height conſiſts of
hands.

The Goths and Cimbri were anciently, like the
Scythians, nomads, and lived alſo like them off
their herds and flocks ; for drink they had pure
water and mead, with mares' milk.[1] This milk,
however, they did not drink unleſs it had firſt
been conſecrated, the horſe being an animal ſacred
to the god of war. Sometimes they drank till
drunkenneſs overcame them of the milk and blood
of their beaſts of burden. They had horſes of
two colours, black and white, and eſteeming one
or the other ſacred, did not ride on both alike.[2]
The beaſts of the ancient Germans, according to
Cæſar, were ſmall and ill-ſhaped, and Tacitus ſays
their horſes were neither conſpicuous for beauty
nor ſpeed, nor were they trained to circle round
at the will of their riders, as were the Roman
cavalry horſes.[3] The Britons attacked in a de-
ſultory way with chariots, now charging their

---

[1] Cnf. Virgil, " Georg.," iii. 463 :

"Et lac concretum cum ſanguine potat equino."

Camilla, the heroïne of the later books of the " Æneid," was
fed as an infant on mare's milk. One of the few babies preſerved by the French through the horrors of the retreat from
Moſcow was kept alive by feeding it on a paſte made of horſe's
blood.

[2] See V. Hehn, *ut ſup*.   [3] Tac., " Germ.," vi.

enemies, now wheeling round their horfes, and again feizing an opportunity when it offered. This was peculiarly annoying to the heavy-armed Roman foldiers, and when the Roman cavalry followed, the chariotmen leapt out and confronted them on foot.[1]

Here the manner in which the horfe was employed in war naturally deferves a word or two. This feems to have been its ufe everywhere before it was utilized for agricultural work, juft as the paftoral ftate of life naturally preceded a more fettled mode of exiftence. So in the Hebrew Scriptures the horfe is exclufively confidered as an animal ufeful in war. Oxen invariably precede it as beafts of draught, juft as we are now feeing it in its turn fuperfeded by fteam. But with regard to the employment of the horfe in war, Lucretius in a celebrated paffage (v. 1296) feems to have mifapprehended its true fequence. " The cuftom of a warrior mounting on horfeback," he fays, " and guiding his fteed with reins and the right hand, is antecedent in time to tempting the dangers of war in a two-horfe chariot; and this, again, to the ufe of four-horfe chariots and chariots armed with fcythes." As a matter of fact, chariots feem to have been ufed before the art of riding on horfeback had been learnt. The Lapithæ were the firft to invent breaking-in of horfes and the ufe of the bridle, while Ericthonius firft introduced the yoking of four horfes to a chariot. The heroes before Troy always fought from chariots, and never

[1] Cæfar, " Bell. Gall.," v. 16.

on horfeback. Grecian and Trojan civilization
as well were juft efcaping, in the ten years' war
before Troy, from thofe facrifices of horfes which,
as we have feen, were wide-fpread in the ancient
world.   It now did not fo much worfhip as
reverence the animal.   "The horfe in Homer
generally has not only a poetical grandeur," fays
Mr. Gladftone, "but a near relation to deity,
which I am unable fufficiently to explain; but
which, it feems poffible, may be the reflection or
analogue of the place affigned to the ox in the
Eaft.   Several circumftances, and among them the
practice of defcribing a champaign country as one
fuited to feeding the horfe, combine to fhow how
completely for the Greek this noble creature ftood at
the head of the animal creation."[1]   While agreeing
in the main with this lover of Homer, we believe
that the femi-divine honour paid to the horfe was
no reflection of ox-worfhip from the Eaft, but an
independent phafe of religious thought.   How
clofely Homer deemed the horfe connected with
the gods, may be feen in the curious narrative of
Hera giving Xanthus, one of the immortal chariot-
horfes of Achilles, powers of fpeech, which the
animal forthwith ufed to foretell its mafter's
fpeedy death.[2]

The ancients chiefly knew of herds of wild
horfes about the river Hypanis[3] and in the vaft
tract which they termed Scythia, which anfwers to
the South of Ruffia, Turkeftan, and the deferts

[1] Gladftone, "Juventus Mundi," p. 360.
[2] "Iliad," xix. 400 *feq.*   [3] Herod., iv. 52.

ftretching into Mid Afia. Palæontology fhews that horfes once abounded in the New World, but thofe which are now found there in a wild ftate are all of them the defcendants of the horfes imported by the Spanifh conquerors, the original horfes of the country having everywhere died out before the introduction of man into the Continent. The aborigines, whom the Spanifh found dwelling in Mexico and Peru, had no tradition or hieroglyphic indicative of fuch a quadruped, and the horfes brought acrofs the Atlantic by the invaders were viewed with aftonifhment and alarm.[1]

At the fiege of Troy, Priam's horfes had been reared at Abydos, which was famous for them. Homer calls Ilios "bleffed with good horfes," and fpeaks of the "horfe-fubduing Trojans" as if they were an equeftrian people. Myths connected Troy with horfes from the beginning, indicative, perhaps, of a Phœnician founder (juft as the emblem of Carthage was a horfe[2]), and thefe legends have been very ufeful to the poets. Thus Zeus gave Tros, the eponymus of Troy, divine horfes in payment for his fon Ganymede, carried off to be a celeftial cupbearer; and Hercules refcued the daughter of Laomedon, Hefione, from a fea-monfter fent by Pofeidon, on the faith

---

[1] Owen, "Hiftory of Britifh Foffil Mammals," p. 398 : "The horfe in its ancient diftribution over both hemifpheres of the globe refembled the maftodon, and appears to have become extinct in North America at the fame time with the *m. giganteus*, and in South America with the *megatherium.*"

[2] " Signum quod regia Juno
Monftrârat, caput acris equi."—(" Æn.," i. 443.)

of a promife that the King would give him fome
fteeds of this divine ftock.   Laomedon, however,
broke his word, and the hero befieged and took
Troy.   During the fecond and more celebrated
ten years' fiege of the city, Æneas poffeffed horfes
of this celeftial ftrain, "the beft of all horfes feen
by the dawn and the fun" ("Iliad," v. 265).
Circe craftily ftole this ftock, and fo their defcendants are faid to "breathe fire from their
noftrils" ("Æneid," vii. 281).   Very fitly, too,
was the deftruction of Troy accomplifhed by the
aid of the wooden horfe, "inftar montis equum"
("Æneid," ii. 15).   Laocoon's advice, fo ill-ftarred
for himfelf, deferves quotation in the original, as
fifty are familiar with the proverb, for one who
knows whence it comes:

> "Equo ne credite, Teucri,
>   Quicquid id eft timeo Danaos et dona ferentes."[1]

The poetic inftinct of Homer compares Paris,
one of the leading champions of Troy, going
forth to battle, to an exulting horfe:

> "As when fome ftall-fed horfe his barley leaves
> And breaks his bonds and clatters o'er the plain,
> Wont there to bathe within the fair-glowing ftream,
> Exulting ; high he bears his head, his mane
> Toffes athwart his neck, and winged with pride,
> Welcomes with lofty fteps the well-known meads."[2]

When Zeus goes forth from Olympus his
horfes are "fwift-flying" in the "Iliad" (viii.
41-43), "their flowing manes tied up with gold."
Milton, on the other hand, when "the chariot of

---

[1] "Æneid," ii. 49.   [2] "Iliad," vi. 506-511.

Paternal Deity" proceeds to war, is ſilent (ſeeing how little ſpace in Hebrew hiſtory the horſe filled) about ſteeds to draw it, but incorporates the grandeſt imagery of the prophets into one of his nobleſt deſcriptions.[1] The two horſes of Achilles's chariot, Xanthus and Balius, flew, ſays Homer, like the wind, and (in accordance with a ſuperſtition common throughout the ancient world) were begotten of the weſt wind, he adds, on Podarga, as ſhe was feeding in a meadow by the ocean ſtream.[2] Similarly Mars has two horſes in his chariot in the " Iliad," named Fear and Terror, though at other times theſe are called his ſons. Homer repreſents Erichthonius as poſſeſſing 3,000 horſes, and 12 foals of marvellous properties, able to run over the ears of corn or the waves without injuring or ſinking in them, were born of theſe by Boreas (" Iliad," xx. 219). Four horſes were ſlain at the pyre of Patroclus, and the reſt of the warriors' chargers were led round the dead body in a rite called by the Romans "decurſio" ("Iliad," xxiii. 10). Horſes were caſt alive into the Scamander to propitiate the river (" Iliad," xx. 130). Bochart in his " Hierozoicon" treats of ancient horſes at large.

Among the Anglo-Saxons no heathen prieſt was allowed to ride on a male horſe (Bede's " Eccl. Hiſt.," ii. 13). None of the moſt ancient gods of Greece were imagined as riding on horſeback. Zeus, Apollo, and the reſt have two-horſe chariots. It is Dionyſos, belonging to a different order of

---

[1] "Paradiſe Loſt," book vi.   [2] " Iliad," xvi. 149.

deities, who firft rides a panther, as Silenus, an afs. Heroes, fuch as Perfeus, Thefeus, and the Diofcuri, are mounted on horfes. Okeanus beftrides a winged fteed (" Prom. Vinct.," 395). The northern gods generally ride; Odin on Sleipnir. He faddles it for himfelf. Night had a fteed Hrîmfaxi (rimy-mane), as Day had Skînfaxi (fhining-mane).[1] In the earlieft period of Teutonic mythology the horfe feems to have been the favourite animal for facrifice. There is no doubt that before the intro-duction of Chriftianity its flefh was conftantly eaten. Nothing in the ways of the heathen was fo offenfive to the new converts as the not giving up the killing of horfes and eating of their flefh.[2] Cæcina, on approaching the fcene of the over-throw of Varus, faw horfes' heads faftened to the ftems of trees. Thefe were the Roman horfes which had been offered up to the German gods.[3]

The Roman "horfey" man ufed to fwear by Hippona, a goddefs of horfes. His Greek equiva-lent appears at the beginning of the "Clouds" of Ariftophanes. The horfe, like the camel, is not found on the moft ancient Egyptian monuments: "Tout ce qu'il eft prudent d'en conclure, c'eft que ces animaux n'étaient ni l'un ni l'autre abondants en Egypte du temps de l'ancien empire, et qu'ils n'étaient point encore comptés alors en nombre des animaux domeftiques" (Chabas, p. 423).

Such are fome of the great affociations connected with horfes in heroic days. Argos, Epidaurus,

[1] Grimm's "Northern Mythol.," ed. Stallybrafs, i. 328.
[2] *Ibid.*, p. 47.   [3] Tac., "Ann.," i. 61.

and Epirus were noted among the Greeks for good horſes. Hence the alluſions, "aptum Argos equis" (Horace's "Odes," i. 7, 9); "domitrix Epidaurus equarum" ("Georg.," iii. 44); "Eliadum palmas Epirus equarum" ("Georg.," i. 59). Wonderful ſtories are told by the ancients of Bucephalus, the horſe of Alexander the Great—how he would allow no one elſe to mount him when harneſſed for war, and when he received his death-ſtroke in a ſkirmiſh in the Indian War, he bore his maſter ſafely out of the battle, and then, and not till then, expired, and the like.[1] A few more notices of famous mythical horſes may be ſubjoined, ſuch as the brazen-footed, fire-ſnorting horſes of Æetes, which it was needful that he who would bear off the Golden Fleece ſhould yoke to a plough, and compel to work. Pegaſus need only be named. Caſtor and Pollux had a celebrated horſe called Cyllarus. On a coin of Rhegium they are both repreſented mounted on him, much like the Knights Templars of later times. The chariot-horſes of Glaucus were a cauſe of ſhuddering to the ancients, as they had gone mad, and torn their maſter limb from limb.[2] Cn. Seius poſſeſſed a horſe of remarkable beauty, ſaid to have ſprung from the ſteeds of Diomedes, whom Hercules had ſlain and brought his horſes from Thrace to Argos, far ſurpaſſing all other horſes in good qualities. Unluckily fate had decreed that everyone who ſhould own it, together with all his houſe, family, and fortune, would be irretrievably ruined. Seius himſelf was

[1] Aul. Gell., v. 2 ; Pliny, viii. 42.   [2] "Georg.," iii. 267.

capitally punifhed by Antony the triumvir. Dola-
bella then fell in love with the horfe, and bought
it for a large fum, but was flain in civil war in
Syria. Caffius was its next owner, and he, on the
rout of his party, put an end to himfelf. Antony
then became poffeffed of it, and his miferable end
it is needlefs to mention. Hence, fays Aulus
Gellius, arofe a proverb of men noted for their
misfortunes—"He owns a Seian horfe."[1] Moralifts
might apply this ftory to the ruin which fo often
overtakes men in modern times who devote themfelves to racing, more efpecially as the horfe of
Seius is defcribed as having been of a dark colour ;
and in the perfon of Pheidippides, the horfe-lover
portrayed in the beginning of the " Clouds " of
Ariftophanes, might defcry the type of many a
" horfey" man of our own times. If horfes were
facred to Neptune, none might ever be brought
near a temple or grove facred to Diana, becaufe
horfes had caufed the death of her favourite, Hip-
polytus.[2]

In our own land the horfe is found on a coin
of Verulamium, the capital of Caffivelaunus. Indeed, it has been noticed that the horfe was a
favourite animal with the Kelts, and that both on
the famous White Horfe of the Berkfhire Downs
and on coins the animal is reprefented with the
wrong leg foremoft in an impoffible attitude. It
was the enfign alfo of the Saxons; but with them
the leg is always correctly drawn (see *Blackwood's
Magazine*, September, 1883, p. 321). A curious

[1] Aul. Gell., iii. 9.   [2] " Æn.," vii. 778.

instance of the use to which its teeth might be put may be seen in the Gibbs's bequest at the South Kensington Museum, where a set of sixty-three draught-men occurs, which date from Anglo-Saxon times. Turning to the fatherland of these Teutonic invaders, it is impossible to forget Odin's celebrated eight-footed horse, Sleipnir. The horse was much offered in sacrifice, and also eaten among the northern nations, before the introduction of Christianity, and there are many indications that the early converts could not wholly give up the eating of horse-flesh. The ancient Germans, after the sacrifice of horses, commonly cut off their heads, and fixed them in some sacred grove as acceptable offerings to their gods.

At the New Year's festival horses were specially sacrificed. We have seen in the more retired districts of Glamorganshire the head of a horse carried round the country at Christmas-time with singing and merriment, which is without doubt a relic of these heathenish superstitions. Pope Gregory III. wrote to St. Boniface so late as A.D. 751, "Among other things, you add that some are wont to eat wild horses, and very many domestic horses: this you should never suffer to be done. Some fowls also, such as jackdaws, rooks, and storks, are to be wholly interdicted from the meals of Christians; beavers also, and hares, and much more wild horses, are to be avoided."[1] Horse-flesh and that of cats

---

[1] "Inter cetera agrestem caballum aliquantos comedere adjunxisti, plerosque et domesticum ; hoc nequaquam fieri deinceps finas. Imprimis de volatilibus, id est graculis et

are more than once named as the food of heathens and witches in northern literature. A curious verſe, which is part of the grace before meat of the monks of St. Gall, points to the uſe of horſe-fleſh ſo late as A.D. 1000—

"Sit feralis equi caro dulcis in hac cruce Chriſti,"

while Profeſſor R. Smith has ſurmiſed that "our own prejudice againſt horſe-fleſh is a relic of an old eccleſiaſtical prohibition framed at the time when the eating of ſuch food was an act of wor-ſhip to Odin."[1] Hippophagy has aſſumed con-ſiderable proportions in Paris of late years, and the following advertiſement from the *Times* of Sept. 16, 1881, ſhews that the northern nations are ſtill true to their old attachment: "Horſe-Fleſh for Exportation.—Wanted, ſound prime Salted Meat in large pieces, ſuitable for ſmoking. Deliveries monthly of about 25 barrels of 200 lb. to 300 lb. each. State price, including packages. f. o. b. London, Liverpool, or Hull. J. C. S——, Landemarket, Copenhagen."

Having thus brought ancient and modern times into juxtapoſition, it is well to remember the poet's line—

"Et jam tempus equûm fumantia ſolvere colla."[3]

---

corniculis atque ciconiis quæ omnino cavendæ ſunt ab eſu Chriſtianorum ; etiam et fibri, et lepores, et equi ſilvatici multo amplius vitandi." See Victor Hehn, *ut ſup.*, p. 24, and Grimm's "Teutonic Mythology," ed. Stallybraſs, vol. i., p. 47, 1880.

[1] "Lectures on the Old Teſtament," p. 366.
[2] "Georg.," ii. 542.

CHAPTER IX.

GARDENS.

"OD ALMIGHTY firſt planted a garden, and, indeed, it is the pureſt of human pleaſures; it is the greateſt refreſhment to the ſpirits of man, without which buildings and palaces are but groſs handyworks; and a man ſhall ever ſee, that when ages grow to civility and elegancy, men come to build ſtately, ſooner than to garden finely, as if gardening were the greater perfection." Thus Lord Bacon begins the ſweeteſt of his eſſays, from every line of which breathe wafts of herbs and flowers. It would be unpardonable, in treating of antiquity, to forget its gardens. Of the original " happy garden," the cradle of mankind, Milton has glorioufly amplified the few outlines traced in the Book of Genefis. Flowers and trees touched his mind almoſt as much as muſic, and he never wearies of dwelling on their beauties. Paradiſe itſelf is twice deſcribed in the great Engliſh epic: once in Book iv. 237-268 ; and again in Book ix. 424-443.

"Eve feparatethe fpies,
Veiled in a cloud of fragrance, where fhe ftood,
Half fpied, fo thick the rofes blufhing round
About her glow'd, oft ftooping to fupport
Each flower of tender ftalk, whofe head, though gay
Carnation, purple, azure, or fpeck'd with gold,
Hung drooping, unfuftain'd ; them fhe upftays
Gently with myrtle band, mindlefs the while
Herfelf, though faireft unfupported flower,
From her beft prop fo far, and ftorm fo nigh !
Nearer he drew, and many a walk traverfed
Of ftatelieft covert, cedar, pine, or palm ;
Then voluble and bold, now hid, now feen,
Among thick-woven arborets, and flow'rs
Imbordered on each bank, the hand of Eve !
Spot more delicious than thofe gardens feigned,
Or of revived Adonis, or renowned
Alcinous, hoft of old Laertes' fon ;
Or that, not myftic, where the fapient king
Held dalliance with his fair Egyptian fpoufe."

Profeffor Heer has refcued fome of the plants and trees which flourifhed in prehiftoric gardens from the buried flora of Switzerland. Such were the following cereals—fmall lake-dwelling wheat, Egyptian wheat, two-rowed wheat, one-rowed wheat, compact fix-rowed barley, fmall fix-rowed barley, common millet and Italian (*fetaria*); peas, poppies, flax, caraway feeds, apples, pears, and bullaces.[1]   The gardens themfelves were probably mere patches of land adjoining caves or lake-dwellings, ufeful for producing corn and a few fruits.

"Retired Leifure,
"That in trim gardens takes his pleafure,"[2]

certainly did not haunt neolithic gardens.

[1] See Dawkins, "Early Man in Britain" (Macmillan, 1880), pp. 301-2.
[2] "Penferofo," 49.

The Affyrians were very fond of formal gardens
fet with trees planted in rows at equal diftances
from each other, and with walks geometrically
regular, efpecially around temples. Canals or
aqueducts frequently fupplied thefe gardens with
water. What Rawlinfon calls "the monftrous
invention of Hanging Gardens,"[1] were known in
Affyria as early as the time of Sennacherib. It was
not till a much later date, however, that they were
introduced into Babylonia, where the celebrated
Hanging Gardens of Babylon were efteemed one
of the wonders of the ancient world. To us thefe
gardens feem rather a laudable attempt to make
the defert rejoice and bloffom as the rofe. They
were conftructed by Nebuchadnezzar to gratify
the home-fick longings of his favourite wife,
Amyitis, and were in the form of " a fquare, each
fide of which meafured 400 Greek feet. It was
fupported upon feveral tiers of open arches, built
one over the other like the walls of a claffic theatre,
and fuftaining at each ftage or ftory a folid platform, from which the piers of the next tier of
arches rofe. The building towered into the air to
the height of at leaft feventy-five feet, and was
covered at the top with a great mafs of earth, in
which there grew not merely flowers and fhrubs,
but trees alfo of the largeft fize. Water was
fupplied from the Euphrates through pipes, and
was raifed, it is faid, by a fcrew, working on the
principle of Archimedes." It was built of bricks,
ftrongly cemented with bitumen, and protected by

[1] Rawlinfon, "Ancient Monarchies," i. 585, and ii. 517.

a layer of sheet lead from the moisture above. "The ascent to the garden was by steps. On the way up among the arches which sustained the building were stately apartments, which must have been pleasant from their coolness. There was also a chamber within the structure containing the machinery by which the water was raised." Professor Rawlinson has put together in these few sentences a mass of information from different classical authorities.

Turning to some of the celebrated gardens of the ancients, partly mythical, partly proverbial, we come first to the Gardens of Adonis, which partook of both these characters. The myth belongs originally to Phœnicia; and the story of Adonis, the favourite of Venus, killed while hunting, and allowed to spend six months alternately with Proserpine and Venus, points not obscurely to the return of summer after winter. Hence "the Gardens of Adonis" is only a poetical expression for summer flowers, and soon passed into a proverb intimating short-lived pleasures. At Athens, the term was used of small pots in which cress and such-like quick-growing herbs were raised. So Plato makes Socrates ask whether any husbandman of sense would wish to see his seeds spring up and flourish with a brief eight-day life in Gardens of Adonis, or would leave them to children and the decoration of feasts, and would sow at the fitting time and be contented if, at the end of eight months, he received his harvest.[1] The Gardens of

[1] "Phædrus," 276 B.

the Hesperides were almost equally celebrated.
Turner has painted them, and Milton spread the
appropriate mist of poetry over these Μακάρων
νῆσοι,

> " Happy isles,
> Like those Hesperian gardens famed of old,
> Fortunate fields, and groves, and flowery vales,
> Thrice happy isles !"[1]

and amplified them in the beautiful imagery of
" Comus," 980-1011.

The Gardens of Alcinous are another proverbial
Paradise. Alcinous was the just and rich King of
the Phæacians in Corcyra, devoted to gardening.
" Quid bifera Alcinoi laudem pomaria?" says
Statius;[2] while " to give apples to Alcinous " was
much like sending coals to Newcastle with us.
Virgil uses these gardens as a synonym for
orchards on account of the fruit which Alcinous
grew, " pomaque et Alcinoi silvæ " (" Georg.," ii.
87). All the Latin poets drew their allusions to
these gardens from Homer. We extract his
account of them from the excellent translation of
the " Odyssey " by Butcher and Lang (" Odyssey,"
vii. 112-131). Thus the reader obtains a literal
rendering free from such verbiage as Pope flings
over the passage : " The reddening apple ripens
here to gold ;" or " Here the blue fig with luscious
juice o'erflows;" and the like. " Without the
courtyard, hard by the door, is a great garden of
four plough-gates, and a hedge runs round on
either side. And there grow tall trees blossoming,
pear-trees and pomegranates, and apple-trees with

---

[1] " Par. Lost," iii. 567.   [2] " Silv.," i. 3, 81.

bright fruit, and fweet figs, and olives in their
bloom.   The fruit of thefe trees never perifheth,
winter or fummer, enduring all the year through.
Evermore the weft wind blowing brings fome
fruits to birth and ripens others.   Pear upon pear
waxes old, and apple on apple—yea, and clufter
ripens upon clufter of the grape, and fig upon fig.
There, too, hath he a fruitful vineyard planted,
whereof the one part is being dried by the heat, a
funny fpot on level ground, while other grapes
men are gathering, and yet others they are tread-
ing in the wine-prefs.   In the foremoft row are
unripe grapes that caft the bloffom, and others
there be that are growing black to vintageing.
There, too, fkirting the furtheft line, are all manner
of garden-beds, planted trimly, that are frefh continually; and therein are two fountains of water,
whereof one fcatters his ftreams all about the
garden, and the other runs over againft it, beneath
the threfhold of the courtyard, and iffues by the
lofty houfe, and thence did the townsfolk draw
water."

Penelope had a " garden of trees " (" Odyffey,"
iv. 737).   Onions, as a relifh for wine, and
poppies were alfo grown in the Homeric gardens
(" Iliad," xi. 629 and viii. 306).   But early Greek
gardens, as a rule, held little but vines and trees,
and were formal in arrangement (" Iliad," v. 90).
The τέμενος, or facred enclofure, often round a
temple, furnifhed a model.   It was planted with
fhrubs, vines, and herbs, and was fometimes termed
ὄρχατος, whence comes our " orchard " (" Iliad,"

vi. 195 ; " Odyſſey," xx. 278), if that word be not rather derived from the A.S. *ortegeard*, or garden of herbs.[1] Laertes, in the " Odyſſey," is repreſented as a gardener, and Ulyſſes on his return finds him " alone in the terraced vineyard, digging about a plant." The ſon addreſſes him, " Old man, thou haſt no lack of ſkill in tending a garden; lo, thou careſt well for all, nor is there aught whatſoever, either plant, or fig-tree, or vine, or olive, or pear, or garden-bed in all the cloſe that is not well ſeen to." Theſe words give ſome idea of a Homeric garden. The pathetic lines of Ulyſſes when diſcovering himſelf to his aſtoniſhed father will fill up ſome of the outlines: " Come, and I will tell thee the trees through all the terraced garden, which thou gaveſt me once for mine own ; and I was aſking thee this and that, being but a little child, and following thee through the garden. Through theſe very trees we were going, and thou didſt tell me the names of each of them. Pear-trees thirteen thou gaveſt me, and ten apple-trees, and figs two ſcore, and as we went thou didſt name the fifty rows of vines thou wouldeſt give me, whereof each one ripened at divers times, with all manner of cluſters on their boughs, when the ſeaſons of Zeus wrought mightily on them from on high.[2]

Roman gardens, again, were for the moſt part formal pleaſure-grounds planted with fruits and

---

[1] For Greek gardens, ſee Becker's " Charicles," p. 203, note (ed. 1880).

[2] " Odyſſey," xxiv. 244 and 335 (Butcher and Lang's tranſlation).

flowers, efpecially fuch flowers as were ufeful for garlands. Both Romans and Greeks too, it fhould be remembered, poffeffed but a limited flora. Our own garden-treafures have been lovingly brought together, carefully cultivated and improved from every clime. What our natural poverty herein would be, may be imagined by mentally excluding all fave native fpecies from our parterres. Lines of trees in a Roman garden bordered ftraight walks laid out for exercife; while fhrubs were cut and trimmed to improve upon nature. Rofes and violets, narciffus, poppy, and a few others furnifhed the borders with flowers. The fecondary pleafures of beauty and natural adaptivenefs of form and growth, which we dwell upon fo largely in our eftimation of a garden, were nearly unknown to the ancients. So Rufkin fuggeftively writes: "I do not know that of the expreffions of affection towards external Nature to be found among heathen writers, there are any of which the leading thought leans not towards the fenfual parts of her. Her beneficence they fought, and her power they fhunned; her teaching through both they under-ftood never. The pleafant influences of foft winds and ringing ftreamlets, and fhady coverts of the violet-couch and plane-tree fhade, they received, perhaps, in a more noble way than we; but they found not anything except fear upon the bare mountain. The Hybla heather they loved more for its fweet hives than its purple hues."[1] Virgil often dwells upon gardens: "Plant now thy pears,

[1] "Modern Painters," vol. ii., p. 17.

Melibæus, plant thy vines in order." "Come hither ; lo, the Nymphs bear thee lilies in brimming baſkets ; for thee a fair Naiad, plucking violets and poppy-heads, twines together the narciſſus and ſweet-ſmelling dill, and twiſting them up with mezereon and other fragrant herbs, varies the delicate hyacinths with yellow marſh-marigold. I myſelf will gather hoary quinces with tender down, and cheſtnuts ſuch as my Amaryllis loved ; I will add waxen plums, honour ſhall alſo be paid to this apple ; you, too, laurels, will I pluck, and you, neighbouring myrtle, ſince thus arranged ye mingle pleaſant odours."[1]   Again, he ſays in the " Georgics," " Let gardens breathing with crocus-flowers invite bees, and the protection of Priapus, that guard of thieves and birds, with his willow cudgel protect them."   The claſſic reader will recall many a ribald ode to Priapus, whoſe image was generally ſet up in gardens.   Three lines in the ſame poem aptly deſcribe a Roman garden :

> " Hæc circum caſiæ virides, et olentia late
> Serpylla, et graviter ſpirantis copia thymbræ
> Floreat, irriguumque bibant violaria fontem."

Dryden muſt tranſlate the moſt celebrated paſſages on ancient flowers (" Georg.," iv. 116-146). He revels in the roſes of Pæſtum, " and their double ſpring," in ſuccory, parſley, cucumbers, narciſſus, bears'-foot, myrtles and ivy, apples, limes, pines and vines, and then deſcribes the old Corycian gardener :

---

[1] " Ecls.," i. 73 and ii. 45-58 ; " Georg.," iv. 109, 30-32, eſpecially 116-146.   A good deal about Roman gardens may be found in Becker's " Gallus."

" Lord of few acres, and thofe barren too,
  Yet labouring well his little fpot of ground,
  Some fcattering pot-herbs here and there he found ;
  Which cultivated with his daily care,
  And bruifed with vervain, were his daily fare.
  For every bloom his trees in fpring afford,
  An autumn apple was by tale reftored.
  He knew to rank his elms in even rows,
  For fruit the grafted pear-tree to difpofe,
  And tame to plums the fournefs of the floes.
  With fpreading planes he made a cool retreat
  To fhade good fellows from the fummer's heat.
  Sometimes white lilies did their leaves afford,
  With wholefome poppy-flow'rs to mend his homely board.
  For late returning home he fupped at eafe,
  And wifely deemed the wealth of monarchs lefs ;
  The little of his own, becaufe his own, did pleafe."

Another enumeration of garden flowers, as
prettily arranged as any nofegay, will be found in
the laft twenty lines of Virgil's " Culex," if that
poem be his, and not merely a monkifh cento.

Having fpoken of prehiftoric gardens, it would
be unpardonable to forget the Egyptian kitchen-
gardens, wherein grew the leeks, onions, and
cucumbers for which the Ifraelites longed. The
fertility of thefe gardens was due then, as now, to
their proximity to the beneficent waters of the
Nile and the alluvial foil of which they were compofed. The celebrated Perfian paradifes were not
gardens at all, but rather parks planted with knots
of trees, wherein fheltered wild beafts until it
pleafed their owners to chafe them. The " terai "
on the flopes of the Himalayas at prefent forms a
good natural example of a paradife. We have mentioned the Saxon " wort-yard," and it is worth remarking that the South of England poffeffed many
vineyards before the Conqueft, though their

grapes would not probably be highly prized at preſent.[1] Every monaſtery and convent would have its own patch of garden ground, and horticultural ſcience in England is largely indebted to the culture and improved varieties of plants introduced by the monks. The celebrated *liqueur* which was recently made by the monks at the Grande Chartreuſe ſhows their ſkill lingering to our own day, as admirably expreſſed by Matthew Arnold:

> "The garden, overgrown—yet mild,
> Thoſe fragrant herbs are flowering there !
> Strong children of the Alpine wild
> Whoſe culture is the brethren's care ;
> Of human taſks their only one,
> And cheerful works beneath the ſun."

There is a Paradyſs (Paradiſe) mead near the Priory of Selborne, Hants, which was probably encloſed ground, planted like an orchard with fruit-trees, and pleaſantly laid out.[2] Jedburgh, in old days, was greatly renowned for pears ; while Buckfaſtleigh is ſaid to have firſt introduced the apple to Devon, owing to the monks at theſe religious houſes having originally planted orchards.

Burton[3] does not forget to eulogize the delights of gardens : " To walk amongſt orchards, gardens, bowers, mounds, and arbours, artificial wilderneſſes, green thickets, arches, groves, lawns, rivulets, fountains, and ſuch-like pleaſant places, like that Antiochan Daphne, brooks, pools, fiſh-ponds, betwixt wood and water, in a fair meadow,

---

[1] See Lappenburg, "Hiſtory of England," ii. 359.
[2] White, "Antiquities of Selborne," Letter 25.
[3] "Anatomy of Melancholy," ed. 1826, vol. i., p. 407.

by a river-fide, muft needs be a delectable recrea-
tion." And he names the prince's garden at
Ferrara, Fontainebleau, "the Pope's Belvedere in
Rome, as pleafing as thofe *horti penfiles* in Babylon,
or that Indian king's delightfome gardens in Ælian,
or thofe famous gardens of the Lord Cantelow in
France." Many of thefe wonders have been
eclipfed by modern marvels of greenery; and fuch
lordly gardens as thofe at Trentham, Chatfworth,
Alton, and others, need fear no comparifon with
any predeceffors. And as for botanical gardens,
our own at Oxford may be worthily matched with
thofe at Nuremberg, Montpellier, or Leyden.

From Saxondom to Chaucer is a long leap, but
the fcantinefs of chronicles, and the little leifure
granted men for gardening in the intermediate
ages, compel us to take it. With his pure love
for flowers and the country, Chaucer delights to
dwell upon the gardens of his time. Thus, in
the "Romaunt of the Rofe," is a garden, lying
four-fquare, enclofed within walls "inftede of
hegge":

> "The gardin was not daungerous
> To herborowe birdes many one;
> So riche a yere was nevir none
> Of birdis fong and branchis grene,
> Therin were birdis mo, I wene,
> Than ben in all the relme of Fraunce."

It is worth while recounting the ordinary
furniture of this garden, as may be gathered
further on in the poem. Ordinary trees were
"laureres, pine-trees, cedres, oliveres, elmis grete
and ftrong, maplis, afhe, oke, afpe, planis long,"

> "Fine ewe, popler, and lindis faire,
> And othir trees full many a paire."

Of fruit-trees appear "pomgranetts a full grete dele," "nutmeggis," "almandris," "figgis, and many a date tre."

To fay nothing of the cedars, the nutmegs here fhow that, poet-like, Chaucer was drawing on his imagination, and that the lift cannot be accepted implicitly as being the contents of a fourteenth-century garden. The next lines, the fpices it contained, prove this more conclufively—"clowe, gilofre, licorice, gingeber, grein de Paris" (grains of Paradife), "canell" (cinnamon), "fetewale of pris" (valerian).

> "And many homely trees there were
> That peches, coines" (quinces), "and apples bere,
> Medlers, plommis, peris, chefteinis,
> Cherife, of whiche many one faine is,
> Notis and aleis" (alife), "and bolas,
> That for to fene it was folas,
> With many high laurer and pine,
> Was rengid clene all that gardine
> With cipris and with oliveris,
> Of which that nigh no plenty here is."

If this garden had no exiftence in the outer world, it at all events fhows what the ideal of a garden was in Chaucer's time—"the platform of a princely garden," as Bacon fays. In "The Pardonere and Tapftere," however, we do get fome idea of what a garden of herbs was like in the poet's day. Therein, he fays:

> "Many a herb grewe for fewe and furgery,
> And all the aleys feir, and parid, and raylid, and ymakid,
> The favige and the ifope yfrethid and yftakid,
> And other beddis by and by frefh ydight."

In the " Affemble of Foules " the poet paints
another delightful garden :

> " A gardein fawe I full of bloffomed bowis
> Upon a rivir in a grene mede,
> There as fweteneffè evirmore inough is
> With flowris white and blewe, yelowe and rede.
> And colde and clere welleftremis nothyng dede,
> That fwommin full of fmalè fifhis light,
> With finnis rede and fcalis filvir bright."

Yet a third exquifitely drawn garden will be
found in " The Frankleine's Tale," " of fwiche
pris," as if it were " the veray Paradis ;" and one
more in " The Merchant's Second Tale " :

> " This gardeyn is evir grene and full of May flowris,
> Of rede, white, and blew, and other frefh colouris,
> The wich ben fo redolent and fentyn fo about,
> That he muft be right lewde therin fhuld route."

The beginning of the " Complaint of the
Blacke Knight " fhould alfo be read by all defirous
of realizing what the gardens of the time refembled. This account of the garden's greenery
contains at leaft one touch that fhould be remembered by lovers of the country:

> " There fawe I growing eke the frefhe hauthorne
> In white motley, that fo fote doth yfmell."

The gardens attached to many of the Middle-
Age caftles are of great intereft. A good example
may be feen at Stirling, of which the charac-
teriftics are the frowning walls of the caftle fur-
rounding it, the little peep at the fky which it
afforded, the fmall fcope there was for a few
bufhes and perhaps a low tree or two to be culti-

vated in it. In the "Knighte's Tale" Palæmon
ſees his Emilia for the firſt time in ſuch a garden :

> "Thurgh a window thikke of many a barre
> Of yren gret, and ſquare as any ſparre."

The ſtory of the Earl of Surrey and the fair
Geraldine may illuſtrate how frequently, in the
immured life which many noble damſels muſt
neceſſarily have led in troublous times, ſuch examples of love at firſt ſight muſt have occurred.

A change has come over the Engliſh garden in
Elizabeth's reign. It contains more herbs and
flowers, and is more daintily laid out, until it
reſembles

> "A paradiſe of delight, to which compared
> Theſſalian Tempe, or that garden where
> Venus with her revived Adonis ſpend
> Their pleaſant hours."[1]

The poets now begin to laviſh ſentiment upon
it ; as, for inſtance, Shakeſpeare, from whoſe plays
a charming Old Engliſh garden can be conſtructed. Richard Barnfield thus enumerates in
1594 the contents of a garden :

> "Nay, more than this, I have a garden plot
>   Wherein there wants nor hearbs, nor roots, nor flowers,—
> Flowers to ſmell, roots to eate, hearbs for the pot,—
>   And dainty ſhelters when the welkin lowers :
> Sweet-ſmelling beds of lillies and of roſes,
> Which roſemary banks and lavender encloſes.

> "There growes the gillifloure, the mynt, the dayzie,
>   Both red and white, the blue-eyed violet,
> The purple hyacinth, the ſpyke to pleaſe thee,
>   The ſcarlet-dyde carnation bleeding yet.
> The ſage, the ſavery, and ſweet margerum,
> Iſop, tyme, and eye-bright, good for the blinde and dumbe.

---

[1] Maſſinger, "Believe as You Liſt."

" The pinke, the primrofe, cowflip, and daffadilly,
  The hare-bell blue, the crimfon cullumbine,
Sage, lettis, parfley, and the milke-white lilly,
  The rofe and fpeckled flower cald fops-in-wine :
Fine pretie king-cups and the yellow bootes
That growes by rivers and by fhallow brookes.

" And many thoufand moe I cannot name
  Of hearbs and flowers that in garden grow."[1]

The *ars topiaria*, which cuts box, yews, hollies,
and the like into the femblance of peacocks or
grotefque monfters, is ufually regarded as the
main feature of the love for gardening which fet
in after the Reftoration, but in truth it was but
the revival of a Roman cuftom. *Topiarius* is the
only name by which an ornamental gardener was
known in good Latin authors.[2]   Pliny fays that
the cyprefs, with its fmall tender evergreen leaf,
readily lent itfelf to the defigns of this functionary,
whether it was required to reprefent hunting-
fcenes or fleets. The periwinkle's evergreen trailers
were alfo preffed into his fervice. Similarly the
acanthus was a " *topiaria et urbana herba.*" Thefe
citations fhow that we have adopted a part for the
whole of what was anciently the topiarian's duty,
viz., the cutting and trimming of fhrubs ; and to
this the topiarian art is now confined.   Pope's
paper in the *Guardian*, Sept. 29, 1713,[3] at once
fwept away the artificial greeneries then in vogue

---

[1] " The Affectionate Shepherd " (Percy Society; vol. **xx.**,
1846, p. 12).
[2] See " Dict. of Greek and Roman Antiq.," *fub voc.*,
" Hortus," and references there.
[3] " On the Art of Gardening," p. 61, by Mrs. Fofter
(Satchell, 1881).

in gardening, and a more natural taſte revived. Then came the era of the landſcape gardeners— "Capability" Brown and his followers. It is un-neceſſary to follow further the fortunes of the art. Rapin has ſung the garden in Latin and Cowper in Engliſh verſe; while Sir Thomas Browne, in his "Garden of Cyrus," and Evelyn in his "Acetaria" and "Sylva," have left claſſical treatiſes which no lover of a garden can afford to neglect. At preſent we ſee a decided revolt from the tyranny of ribbon-beds, zones of colour, and the frigid artificial ſtyle which has for ſome years found favour with ſociety, to a more natural and leſs laborious character, in which ſimplicity far tranſcends art, in the eyes of all who have ſtudied the relations between theſe two principles of gardening. The effects of geometrical gardening and lines of bedding-plants can be ſeen with more permanence in a brilliant carpet; for the delight-ful reſults of improving Nature and preſſing her wildings into a decent conformity with man's needs and his ſenſe of beauty, we muſt reſort to ſome ſuch charming piece of tutored negligence as was ſo daintily depicted by Lord Beaconsfield in the garden of Coriſande.

CHAPTER X.

HUNTING AMONG THE ANCIENTS.

ἀμφιβαλὼν ἄγει
καὶ θηρῶν ἀγρίων ἔθνη
περιφραδὴς ἀνήρ.

(Soph. *Antig.*, 344.)

ITH man, as among the lower animals, neceſſity led to the practice of hunting. Inſtinct bids them each purſue what it can ſtrike down, kill, and eat.

" Say, will the falcon ſtooping from above,
Smit with her varied plumage, ſpare the dove ?
Admires the jay the inſect's gilded wings ?
Or hears the hawk when Philomela ſings ?"[1]

Hunting is a wide word, and embraces many different quarries. Nimrod was the firſt hunter, and his prey was man. But here hunting will be narrowed to the chaſe of quadrupeds. And Izaak Walton's huntſman ſhall eulogize his favourite ſport : " Hunting is a game for princes and noble perſons ; it hath been greatly prized in all ages ; it was one of the qualifications that Xenophon

[1] Pope, " Eſſay on Man," Ep. 3.

beſtowed on his Cyrus, that he was a hunter of
wild beaſts. Hunting trains up the younger
nobility to the uſe of manly exerciſes in their
riper age. What more manly exerciſe than hunting the wild boar, the ſtag, the buck, the fox, or
the hare? How doth it preſerve health and in-
creaſe ſtrength and activity!" And once more:
" What muſic doth a pack of dogs then make to
any man, whoſe heart and ears are ſo happy as to
be ſet to the tune of ſuch inſtruments!"[1]   When
Jupiter implanted an evil nature in beaſts which
were at firſt harmleſs, ſays the Latin poet:

> " Tam laqueis captare feras et fallere viſco
> Inventum, et magnos canibus circumdare ſaltus."[2]

In the golden age men had no knowledge of
agriculture ; nor were they careful to heap up
riches or to be thrifty in the uſe of what they
poſſeſſed :

> " Sed rami atque aſper victu venatus alebat."[3]

When civilization began, the hunting exiſtence
gave way to the paſtoral ſtate, and that to the
ſettled mode of living implied by the cultivation
of land. And when pleaſure and luxury abound
in a ſtate, men revert for amuſement to what their
anceſtors had been compelled to practiſe from
neceſſity. In old days man hunted for his dinner ;
now he hunts in order to gain an appetite for it.

Horace held in high eſtimation hunting, and the

---

[1] " The Compleat Angler," i. 1.
[2] Virgil, " Georg.," i. 139, 140.
[3] ";Æn.," viii. 318.

leading out of mules laden with Ætolian toils
and dogs into the country was "a work of fpecial
importance to Romans, ufeful for their reputation,
their health, their morals, and the more fo if
you have ftrength enough either to furpafs the
hound in running, or conquer the boar by thews
and finews."[1]   Plato, too, looks with much fondnefs on the chafe.   His model legiflator is to
frame enactments concerning it, "commending
that kind of hunting which will make the fouls
of young men better, and blaming the contrary
kinds."   Fifhing and fowling may be all very well
for their profeffors, but hunting quadrupeds with
horfes and dogs, and fighting them hand to hand
with miffiles, as in a Homeric hunting-piece, is the
only fpecies of hunting which fhould be fuffered
among high-born youths.   Any kind of fetting of
traps or nets, and deceiving the quarry in the dark,
is hateful;[2] but let no one ftop thofe who are in
fober earneft facred huntfmen, wherever and in
whatfoever guife they choofe to hunt.   Fowling
and fifhing are not very noble taftes for any young
man ; they fhould be left to thofe who are compelled to practife thefe crafts in order to earn their
fubfiftence.   With the whole oriental world,
hunting was held in fpecial favour.   Hunting-
pieces conftantly appear in Egyptian imagery.
The Parthians were devoted to the chafe.   The
Affyrian and Babylonian monarchs conftructed
large "paradifes," as the Greeks called them,

---

[1] "Ep.," i. 18, 49.
[2] Plato, "Laws," bk. vii. 823-4.

where wild beaſts found ſafe harbour until it pleaſed their maſters to hold a grand hunting-party and ſlay them. They are deſcribed as having confiſted of ſpacious tracts of grazing-land, with plantations, and woods, and cool ſtreams within them, ſomething like the Terai of Nepaul at the preſent day. Cyrus's whole army, in which Xenophon was ſerving, was re-viewed in one of theſe.[1] The latter wrote a treatiſe on hunting. Varro, Arrian and Julius Pollux give much information on the ſame ſub-ject. Three treatiſes on hunting, fiſhing, and fowl-ing are alſo aſcribed to Oppian. The epitaph on the tomb of Darius ſhews the keenneſs of the Perſians for the chaſe: "I was a friend to friends; I became the moſt ſkilful of horſemen and archers; I was a maſter in the art of hunting; I could do all things."[2] When Paulus Æmilius ſubdued Macedonia, he is ſaid to have brought away the hounds and hunting-eſtabliſhment of Perſeus, the conquered king, to Rome, and given them to his ſon Scipio Æmilianus. With the Germans, again, "their whole life was ſpent in hunting and the ſtudies of warfare," ſays Cæſar.[3] In our own time theſe have been the only reſources of the North American Indians. Fighting and hunting all over the world form the amuſements of every vigorous race in the infancy of civilization.

---

[1] See Lord Cockburn's article on "Ancient Hunting," in the *Nineteenth Century*, October, 1880; and for ancient authorities, Kreyſig G. C., "Bibliotheca Scriptorum Venaticorum." 1750, 8vo, Altenburgi.

[2] Strabo, xv. 3, 8.   [3] "De Bell. Gall.," vi. 21.

Homer celebrates Scamandrius as an early
hunter, "for Artemis herſelf taught him to hurl
his darts at all the wild monſters which the wood on
the mountains nouriſhes." So Virgil's Lauſus was
"equûm domitor debellatorque ferarum."[1] Many
beautiful hunting-pictures may be found in Homer,
and from no ſubject ſo frequently as the chaſe,
are the ſimiles in the "Iliad" drawn. Lion and
wild-boar hunting are ſpecially dear to Homer.
Here is a ſpecimen: "As when among dogs and
hunters a wild boar or lion turns hither and
thither, rejoicing in his ſtrength, and they, having
drawn themſelves up tower-wiſe, ſtand oppoſite it
and hurl from their hands many javelins, but its
ſtout heart never quails or dreads, and its own
nobility proves its death."[2] The dogs were taught
to ſeize theſe animals from behind, and "truſted
in their ſwift feet." The hunters cheered on their
hounds. Here is another picture which reminds
us of Snyders's hunting-pieces: "As when hounds
and impetuous youths purſue a wild boar, and he
breaks covert from the thick bruſhwood, ſharpen-
ing his gleaming tuſk with crooked jaws; around
him they preſs, but low down comes the gnaſhing
of his tuſks, and they await his charge, dreadful
though he be, ſo," etc. Again: "But they, as
when dogs and ruſtics have chaſed a ſtag with
large antlers, or a boar, and it ſteep rock and
thick coverts have ſheltered, nor is it fated for
them to light upon it, but at their ſhouting a lion

[1] "Iliad," v. 51.
[2] *Ibid.*, xii. 41 ; viii. 338 ; xi. 293 ; xi. 414 ; xv. 271.

with patriarchal mane appears on the road and quickly puts them to flight, eager as they are, ſo," etc. Once more: "But they ruſhed forth and fought before the gates, like wild ſwine which have awaited in the mountains the advancing uproar of men and dogs, and ruſhing ſideways, break up the thicket around them, cutting it up by the roots, and from beneath riſes a gnaſhing of tuſks, until ſome one is ſmitten and loſes his life."[1]

The moſt lifelike of all Homer's hunting-pieces, however, is found in the "Odyſſey." It ſeems to have been ſtudied from an actual occurrence, ſo freſh and animated are the verſes which embalm it. They relate how, in his youth, the hero received the wound on the leg by which, on his return from his twenty years' wandering, his old nurſe Euryclea diſcovered him. "They fared up the ſteep hill of wood-clad Parnaſſus, and quickly they came to the windy hollows. Now the ſun was but juſt ſtriking on the fields, and was come forth from the ſoft flowing ſtream of deep Oceanus. Then the beaters reached a glade of woodland, and before them the hounds ran tracking a ſcent, but behind came the ſons of Autolycus, and among them goodly Odyſſeus followed cloſe on the hounds, ſwaying a long ſpear. Thereby in a

[1] "Iliad," xii. 146. Xenophon in his "Treatiſe on Hunting" ſpeaks but little of hunting ferocious animals. Hare-hunting is his delight. He deſcribes all the knots, ſlips, ſnares, etc., neceſſary for it, with all the detail of accompliſhments and tools ſuited to the mediæval angler. (See Muir's "Literature of Ancient Greece," vol. v., p. 477, etc.)

thick lair was a great boar lying, and through
the coppice the force of the wet winds blew never,
neither did the bright fun light on it with his rays,
nor could the rain pierce through, fo thick it was,
and of fallen leaves there was great plenty therein.
Then the noife of the men's feet, and of the dogs
came upon the boar, as they preffed on in their hunting, and forth from his lair he fprang towards them,
with his back well briftled and fire fhining in his
eyes, and ftood at bay before them all.   Then
Odyffeus was the firft to rufh in, holding his fpear
aloft in his ftrong hand, moft keen to fmite; but
the boar was too quick for him, and ftruck him
above the knee, ripping through much flefh as he
charged fideways, but he reached not to the bone
of the man.   But Odyffeus fmote at his right
fhoulder and hit it, fo that the point of the bright
fpear went clean through, and the boar fell in the
duft with a cry, and his life paffed from him."[1]
This is exactly the place where the "pigfticker"
on the plains of India ftill endeavours to transfix
a wild boar, another proof that the lines may
have been infpired by fome perfonal adventure of
Homer.   The woes of the hunter, "as he ranges
over the peaks of the mountains," are feelingly
dwelt upon by Homer,[2] recalling Horace's "venator
fub Jove frigido."
    A common mode of hunting large animals was
by enclofing them with a ring of men and dogs,
through which it was difficult to break.   "As a

---

[1] "Odyffey," xix. 431-454 (Butcher and Lang).
[2] *Ibid.*, ix. 121.

lion deeply ponders among a crowd of men, in fear when they draw round him the crafty circle," fays Homer.[1]   The cuftom lafted till recent times in Scotland, and the ring thus formed was known as the *Tinchel :*

> " We'll quell the favage mountaineer
> As their Tinchel cows the game."[2]

A peep at the implements of ancient foreft-craft is allowed us in Virgil's celebrated hunting-fcene, when Æneas and Dido went forth together on a fateful morn, " wide-mefhed nets, toils, and boar-fpears with broad fteel heads."[3]   Along with thefe were the *alæ*—coloured feathers faftened on ropes, which were fufpended fo as to allow them to flutter in the wind and terrify the wild creatures till they dafhed into the feries of toils[4] which were fet for their capture :

> " Dum trepidant alæ faltufque indagine cingunt."[5]

In another fplendid paffage, defcribing the manner in which during winter the Northern nations capture deer, the poet again introduces thefe " alæ."   The ftags are found in herds, half fmothered in fnow, which their horns can hardly furmount (fomething in the fafhion of moofe in a Canadian " yard ") ; " thefe they flay with fteel at

---

[1]   " Odyffey," iv. 791.   [2]   " Lady of the Lake," vi. 17.
[3]   " Æneid," iv. 131.
[4]   For the manner in which the toils were fet, compare
> " The toils are pitched and the ftakes are fet,
> Ever fing merrily, merrily."
>   (" Lady of the Lake," iv. 25.)
[5]   " Æneid," iv. 121.

clofe quarters, without flipping hounds at them, without any toils, nor do they terrify the timorous creatures by the fear of the fluttering crimfon feather, as they vainly thruft at the mafs of fnow oppofing efcape, and bellow hoarfely."[1]   Our Englifh Bible, in Ifa. xxiv. 17, reproduces the hunting terms of the Vulgate, "formido et fovea et laqueus," "fear and the pit and the fnare;" but the Hebrew word for "fear" does not feem to have the technical meaning of "formido" as a hunting term.   To return to Dido's hunting; Dryden here rifes to the occafion:

> "Now had they reached the hills and ftormed the feat
> Of favage beafts in dens, their laft retreat;
> The cry purfues the mountain goats, they bound
> From rock to rock and keep the craggy ground.
> Quite otherwife the ftags, a trembling train
> In herds unfingled, fcour the dufty plain,
> And a long chafe in open view maintain.
> The glad Afcanius, as his courfer guides,
> Spurs through the vale, and thefe and thofe outrides,
> His horfe's flanks and fides are forced to feel
> The clanking lafh and goring of the fteel.
> Impatiently he views the feeble prey,
> Wifhing fome nobler beaft to crofs his way;
> And rather would the tufky boar attend,
> Or fee the tawny lion downward bend."[2]

Somerville is diftinctively the poet of the chafe, but his wordy blank verfe does not compare favourably with the vigorous, fwift rhymes of Dryden.   Hear the latter, when untrammeled by

---

[1] "Georg.," iii. 372 : "Puniceæve agitant pavidos formidine pennæ." "Formido" is a technical hunting term fignifying any terror tricked up with feathers.
[2] "Æneid," iv. 151-159.

the needs of tranflation, on the chafe of the king
of beafts :

> " So Libyan huntfmen on fome fandy plain,
> From fhady coverts roufed the lion chafe ;
> The kingly beaft roars out with loud difdain,
> And flowly moves, unknowing to give place.

> " But if fome one approach to dare his force,
> He fwings his tail and fwiftly turns him round,
> With one paw feizes on his trembling horfe,
> And with the other tears him to the ground."[1]

One or two hunting pictures will fhow the felicitous touch of Virgil. To begin with the wild
boar, "animal propter convivia natum" (Juvenal,
i. 141) ; no finer and more lifelike fcene could
be painted than that in " Æneid," x. 708-16 :
"And as that wild boar, driven from the lofty
mountains by the grip of hounds whom pine-
bearing Vefulus has fheltered for many a year,
and many years alfo the Laurentian marfh, having
battened on the tall reeds, after it has come to
the nets, is wont to ftand ftill, and fiercely gnafh
his tufks and fet up his briftling flanks, nor has
anyone the courage to dare his rage or draw near,
but from afar men ply him with darts and fafe
clamours. He then, fearlefs, delays his charge,
firft on one fide, then on another, champing his
teeth, and fhaking off the javelins from his hide,
fo," etc. In contradiftinction to the timidity of
thefe hunters, a coin of Nero fhows a man boldly
confronting a wild boar with a fhort fteel fpear.
In another paffage the chafe of a ftag by an

---

[1] "Annus Mir.," 381.

Umbrian hound, a variety much valued by ancient hunters, is beautifully defcribed :[1]

> " Thus, when a fearful ftag is clofed around
> With crimfon toils, or in a river found,
> High on the bank the deep-mouthed hound appears
> Still opening, following ftill, where'er he fteers ;
> The perfecuted creature, to and fro,
> Turns here and there to 'fcape his Umbrian foe ;
> Steep is th' afcent, and if he gains the land
> The purple death is pitched along the ftrand ;
> His eager foe, determined to the chafe,
> Stretched at his length, gains ground at every pace ;
> Now to his beamy head he makes his way,
> And now he holds, or thinks he holds, his prey ;
> Juft at the pinch the ftag fprings out with fear,
> He bites the wind, and fills his founding jaws with air.
> The rocks, the lakes, the mountains, ring with cries,
> The mortal tumult mounts and thunders in the fkies."
>
> DRYDEN.

The Englifh poet amplifies the original into twice the number of lines, without augmenting in any fenfible degree his Roman brother's imagery ; and careful, and in fome of the lines excellent, as is the tranflation, to feel the more chaftened beauty of the Latin tongue the reader fhould confult the original, which is terfe, vivid, and cumulative in intereft in the higheft degree. Indeed, without reference to Virgil, moft Englifh readers will be hopeleffly confufed about the meaning of Dryden's line :

> "The purple death is pitched along the ftrand."

Another charaƈteriftic defcription, replete with Virgil's ornate tendernefs, might be quoted—the chafe of Silvia's pet ftag by Afcanius and his

---

[1] " Æneid," xii. 749-757. For " puniceæ feptum formidine pennæ," fee note 1, p. 150.

hounds (" Æneid," vii. 483 *ſeq.*) ; but a ſtill more pathetic image—the compariſon of the love-ſick Dido to a ſtricken hind—claims precedence: " The flame of love devours her unreſiſting heart, and the ſilent wound lives and glows beneath her breaſt. Unhappy Dido is conſumed by its fires, and wanders demented through the whole city, as a hind ſmitten by the arrow which from afar, as ſhe incautiouſly roamed the Cretan groves, a ſhepherd has transfixed while ſhooting, and unwittingly left his winged ſteel rankling in the wound. She, in flight, ruſhes through the Dictæan woods and lawns in vain ; the deadly arrow clings to her ſide " (" Æneid," iv. 66).

Hunting chiefly went on in the wintry months with the Romans, as it does with us. " Then," ſays Virgil, " is the time to lay ſnares for cranes, and ſet nets for ſtags, and purſue the long-eared hares ; then, too, ſhould a man whirl round his head the thongs of the Balearic ſling and ſlay the hinds, when the ſnow lies deep, when the rivers ſweep along maſſes of ice " (" Georg.," i. 307). And ſo the love-ſick Gallus ſings: " Meanwhile, together with the Nymphs, I will wander over Mænalus or chaſe fierce boars ; no cold ſhall prevent my ſurrounding the Parthenian groves with my hounds. I ſeem to be hurrying through rocks and reſounding thickets, my delight to wing Cretan ſhafts from a bow tipped with Parthian horn—as if theſe things could avail againſt my madneſs !" (" Eclogues," x. 56). The Romans of later days poſſeſſed at their villas *vivaria* or *leporaria*, which were enclofures holding not merely hares for the purpofe of hunting, but even wild boars. Thefe enclofures (refembling the Perfian paradifes), were also called *roboraria*, from their ftrong oak palings.[1]

Such was the enthufiafm for hunting under the early Cæfars that Juvenal fatirizes the noble matron who with naked breaft like an Amazon meets the rufh of a wild boar and transfixes him with her fpear (i. 22). Nets and enclofures for deer were frequent in England fo early as the Conqueft. Roe-deer were thus taken in Lancafhire, as we learn from "Domefday Book," where a certain Roger had a "haia capreolis capiendis."[2] Mr. Harting explains this paffage thus: "The 'haia,' 'haye,' or 'haie,' as it is varioufly fpelled, properly fignified the hedge or fence enclofing a foreft or park, but by an eafy metonymy the word was transferred from the enclofing fence to the area enclofed by it. In the cafe of the roe-deer it doubtlefs implied an enclofed area, into which the animals were driven, and having outlets here and there acrofs which nets were hung for their entanglement and capture." In the Middle Ages thefe enclofures were called *parci* or *faltatoria*. A ftrong infufion of the hunting element came into England with the Northmen, whofe two chief amufements were fighting and hunting. Everyone will remember

---

[1] See a difcuffion on thefe terms in Aulus Gellius, ii. 20.

[2] See a good paper on the "Roe Deer," *Pop. Science Review*, April, 1881, p. 138.

the famous wild boar of Northern Mythology, Sarhimner, who was hunted every day by the heroes in Walhalla, and feaſted upon every night, and then miraculouſly came to life again next day for another chaſe, thus affording eternal amuſement to his purſuers.

Time would fail to recount the different modes of hunting which ſucceeded the uſe of nets and toils, of the croſsbow-ſhooting and ſlipping of dogs at deer practiſed in and after the Norman period of Engliſh hiſtory. Our purpoſe is but to touch upon the early phaſes of the ſport. He who would know ſomething of hunting in the Middle Ages ſhould conſult the quaint treatiſe on it—earlieſt in the language—in the " Boke of St. Albans," 1486, or the fuller pages of Jacques de Fouilloux, 1650. Beckford and Surtees bring the art of fox-hunting—to which moſt Engliſh hunting has ſhrunk—down to our own days. Vanière, the Jeſuit poet, well deſcribes the moral uſes of hunting:

"Nobilium labor ille virûm eſt, bellique cruenti
Dulce rudimentum ; juvenes exercita curſu
Corpora venando durant ad frigus et æſtum,
Corda ſibi generoſa parant, animamque capacem
Mortis, et expertem media inter tela pavoris ;
Exercent et ad arma manus ; aſtuque ferarum
Ac nemorum inſidiis et bellica furta docentur
Hoſtileſque dolos." [1]

---

[1] " Præd. Ruſticum," lib. xvi.

CHAPTER XI.

MUCH as Great Britain owes to Rome for her gifts of literature and law, civilization was even more largely aided by her in the ordinary conveniences of daily life. The arts of building and road-making among us retain monuments which at the prefent day and for many ages to come will fhow the eminence of the Romans in thefe neceffary arts. Even ftone buildings with windows and chimneys were firft erected by them. It was not probable that Rome, having fucceeded in pacifying the country, fhould not introduce along with improved proceffes of agriculture, and in villas rivalling thofe of Italy or the fouth of France, plants and animals to fill the farmyards and gardens. Indeed, an involuntary introduction of them neceffarily attends any invafion on a large fcale. Weeds and wayfide flowers hitherto unknown to France followed, during the war of 1870-71, the invading footfteps of the Germans.

The lapfe of time, too, during which plants and animals could be brought by the Romans to a friendly population in Britain fhould be noted. The country was fubdued and fettled by Suetonius, Paulinus and Agricola before the end of the firft century after Chrift, and, in fpite of many viciffitudes, the Romans did not finally withdraw from Britain until the beginning of the fifth century, when the affaults of the Goths and the calamities engendered by inteftine wars were rapidly breaking up the Empire. The independence of Britain was declared by Honorius, the lawful Emperor of the Weft, in 410, by his celebrated letter bidding the Britons provide for their own fafety againft the marauding Saxons, Picts, and Scots. But it was not only during this period of Britain's exiftence as a province for fome 400 years that intercourfe with Rome continued. The moral fupremacy of that ancient centre of both mental and material civilization was fully recognifed during the reigns of Saxon and Danifh kings. Save in the mountainous faftneffes, the country was ftudded with temples, bafilicas, baths, and bridges, for which it was indebted to Rome. And yet, however fkilled the Romans were in hufbandry, it is noticeable that all the agricultural implements ufed among the Saxons, which have come down to our days, bear German names. The fame is the cafe with the names of the meafures of land— rood, acre, and the like.[1] The arts and fciences, however, regarded Rome as the centre of infpira-

[1] Lappenberg, "Hiftory of England," vol. ii., p. 359.

tion, until the hordes of Saxon invaders almoſt deſtroyed the old civilization. Thule, according to the ſatiriſt, had engaged her rhetorician in the firſt century, and Britiſh eloquence was not unknown to the, capital of the Weſtern world before the fifth century. Britiſh Chriſtianity more than any other cauſe contributed, after the Romans had left Britain, to the influences of the imperial city being ſtill cheriſhed. Britiſh ſaints and Britiſh heretics alike ſwelled the fame of their country, and promoted direct intercourſe with Rome. The waves of Teutonic deluge ſwept over the land, and well-nigh obliterated all traces of Roman civilization, ſave thoſe which were too maſſive to be readily overthrown. With the coming of Auguſtine, Rome again, and more powerfully than before, becauſe ſhe now ſubjugated the ſouls of the people, reſumed her ſway in humaniſing Britain, and introducing freſh elements of civilization. It becomes an intereſting queſtion, in conſidering the evidences which yet remain of the material conveniences of life which Rome contributed to our land, to determine what plants and animals ſhe brought to our iſland.

The queſtion is complicated by the fact of there having been two other epochs, ſince Roman influences worked, to which the introduction of many plants and animals, now fairly domeſticated among us, may be referred: the return of Weſtern chivalry from the Cruſades and the influx of monks which overſpread Britain during the twelfth and thirteenth centuries. The latter of theſe cauſes was undoubtedly the more conſiderable, and to the Ciſtercians and Benedictine monaſteries, which gradually ſprang up over the country, modern England owes moſt of her ſmaller cultivated plants and fruits. Devonſhire yet recogniſes the parent of her modern cider orchards in Buckland Monachorum, while the pears of Jedburgh are famous in the Border diſtricts. Ages before the coming of the Romans, neolithic man on the Continent, according to Profeſſor Heer, cultivated eight cereals, together with peas, poppies, flax, carawayſeeds, apples, pears, and bullaces; but it is probable that few of theſe latter plants had found their way to Britain until the Roman invaſions. Italy herſelf was very hoſpitable to the animals and trees of the Eaſt, when ſhe ſubjugated one by one its different countries; and as ſhe ſwept into her own boſom all their gems and works of art, ſo was ſhe forward to accept and foſter animals or vegetables which were likely to miniſter to her profit or convenience. Virgil ("Georg.," ii. 64-71) enumerates a few of the latter, Paphian myrtles, the huge aſh, the poplar for making garlands in honour of Hercules; the Grecian oaks, pines, hazels, planes, horſe-cheſtnuts, mountainaſhes, pears, and cherries—even palms, though theſe would not bear fruit in Italy, and were uſeful only for their leaves. In ſhort, Italy was a nurſery in which the plants and fruits of the world were domeſticated.[1] Nothing was more natural than

[1] Columella, iii. 9, 5, "His tamen exemplis nimirum ad-- monemur curæ mortalium obſequentiſſimam eſſe Italiam, quæ

that fhc fhould beftow on the Weft a few of the comforts of life which fhe herfelf had received from the Eaft.

To take thefe in order, we may begin with quadrupeds. In the few touches with which Cæfar paints the fauna and flora of Britain, he fays ("De Bell. Gall.," v. 12), "pecorum magnus numerus;" and further on, "lacte et carne vivunt." Thefe domefticated oxen and fheep were legacies from neolithic man. The oxen in all probability were the anceftors of the few wild cattle at prefent living in Chillingham Park and a few other localities, the *bos taurus* of Linnæus. Another fpecies, the fhort-horned Celtic ox (*b. longifrons*), although its remains have been found in Britain affociated with thofe of the elephant and rhinoceros, was domefticated in England during the Roman period, and fupplied the legionaries with food. It feems likely that our Welfh and Highland cattle, and alfo the red Devon breed, are defcended from it; and that the Romans were either the firft to domefticate it in Britain, or elfe that they intro-duced a better breed than that already in fubjec-tion to the Britons. Owing to the numerous divergent breeds of oxen, at prefent it is difficult to afcertain the original fpecies.

The afs, the mule, and the goat were alfo in-troduced from Rome. Foffil forms of the afs and goat have been found in Britain, but moft naturalifts now believe that our afs is defcended

---

pæne totius orbis fruges adhibito ftudio colonorum ferre didicerit."

from the *equus tæniopus* of Abyſſinia, and our goats from the *capra ægagrus* of the mountains of Aſia, poſſibly mingled with the allied Indian ſpecies *c. falconieri*.[1]   They filtered to us through Roman influences.   The cat alſo came from Rome, as has been ſhewn in another chapter.   The fallow-deer had exiſted in Great Britain in prehiſtoric times, but ſeems to have become extinct before the Roman period, and modern ſo-called "wild" fallow-deer in Scotland are all deſcended from eſcaped ſpecimens, deſcendants of thoſe brought to our iſland from Rome.[2]   This deer is originally an inhabitant of the diſtricts bordering on the Mediterranean.

The rabbit, although often deemed indigenous to Britain, is another native of the Ciſalpine countries of the Mediterranean baſin, and is plentiful in Greece and parts of Italy.[3]   Cæſar ſays that the ancient Britons did not deem it lawful to taſte the fleſh of the hare, the hen, and the gooſe, but ſays nothing of the rabbit.[4]   Martial has a well-known couplet on the animal:

> "Gaudet in effoſſis habitare cuniculus antris,
> Monſtravit tacitas hoſtibus ille vias."[5]

This uſeful (or deſtructive) animal, as the caſe may be, alſo came to us from Rome.   When John, Earl of Morton (Mortain in Normandy),

[1] Darwin, "Domeſtication of Plants and Animals," vol. i., pp. 65 and 105.
[2] Alſton, "Fauna of Scotland," 1880, p. 24 ; and Bell, "Brit. Quads.," ed. 2, p. 358.
[3] Bell, *ut ſup.*, p. 343.   [4] "Bell. Gall.," v. 12.
[5] Martial, xiii. 60.

gave by grant immunities to his tenants outfide
the regard of Dartmoor Foreft, he exprefly allows
them to take the roe, fox, wild-cat, wolf, hare, and
otter, but no mention is made of the rabbit, which,
perhaps, was not common yet in that diftrict. So
in the accounts of Exeter College, Oxford, for
1361, 12d. is charged for four ducks, 11d. for
two fucking-pigs, 1d. for onions, and 8d. for
rabbits.[1] They were probably 4d. or 5d. each, as
they were then fcarce. From us this animal has
found its way with difaftrous confequences to
Auftralafia.

Turning next to birds, we will begin with what
Lucretius beautifully calls—

> "Aurea pavonum ridenti imbuta lepore
> Sæcla."[2]

Peacocks are natives of the Indian jungles; and
fo Curtius, fpeaking of Alexander the Great's expedition, fays: "Thence they marched through a
defert to the river Hydraotes" (now the Ravee in
the Punjaub). "Adjoining it was a wood, gloomy
with trees elfewhere unknown, and filled with a
multitude of wild peacocks." Solomon imported
peacocks from the Indian Ocean. It was a bird
unknown to Homer, and was received by the
Greeks from the Perfians; the Greek, Perfian,
and Hebrew names for the bird being very much
alike. Its feathers were ufed for luxury in Greece
and Rome, and the bird itfelf formed a celebrated
plât at banquets. For this purpofe thofe which

[1] "Regifter of Exeter College," 1879, p. ix (note).
[2] Lucretius, ii. 502.

came from the Ifle of Samos were moft valued.[1]
Horace does not forget the peacock at his feafts,
and Juvenal fatirizes the indigeftion of the glutton
who "carries a whole peacock infide him when he
goes to the bath" (i. 143). This proud bird
was facred to Juno, and is often found on the
coins of the Cæfars as a fign of the "confecratio"
of their female relatives, juft as the eagle pointed
to the apotheofis of the males of that family.
There can be no doubt that our lordly terraces are
indebted to the Romans for their peacocks.[2] The
pheafant is another bird of brilliant plumage
which alfo came to us from Rome. Its home is
Colchis; fo Statius fays to gluttons: "Ah,
miferable men who delight to know how far *the
bird of Phafis* furpaffes the wintry crane of
Rhodope" ("Sylv.," iv. 7). Ariftophanes alfo
tells us that pheafants were dear to gluttons.
Pliny notices that in Colchis the pheafant could
raife and deprefs two earlike feathers. Both
peacock and pheafant were probably brought to
Britain to grace the villas of its Roman conquerors.[3] Guinea-fowls alfo arrived in Britain at
the fame time. They were known as Numidian
or African fowls to the Roman poets, from their

[1] Aulus Gell., vii. 16.

[2] The peacock is mentioned in Chaucer, "Romaunt of the
Rofe;" and Profeffor Rogers thinks it was not introduced until
the thirteenth century (Greenwell and Rollefton's "Britifh
Barrows," 1877, p. 744).

[3] In 1199 a certain W. Brewer was licenfed to have "free-
warren throughout all his own lands for hares, *pheafants*, and
partridges" (Dugdale); fo the pheafant was at that time accli-
matized in Englifh woodlands.

native home. Martial calls them "Numidian spotted fowls," which exactly defcribes their beautiful plumage. Varro, writing fome thirty years B.C., fays that they were the moft recent addition to the glutton's *menu*. Geefe and ducks would naturally be domefticated by the Britons, as foon as they fettled down into an agricultural life, from the wild fpecies, but improved varieties were brought over by the Romans. The turtle-dove, a native of India, is faid to be another introduction. The jungle-cock of the Indian forefts had already made its appearance together with the neolithic man in Britain. The ufe of the falcon, too, in fowling has with fome probability, feeing how popular is falconry in Perfia and the Eaft, been afcribed to the Romans, from whom our anceftors would learn it, and then excel their teachers with native birds.

The trees and vegetables which have been introduced from the Miftrefs of the World open a much larger queftion. When the curtain rifes upon our ifland and hiftory begins, Cæfar obferves upon its vegetation: "Materia cujufque generis, ut in Gallia eft, præter fagum atque abietem" ("Bello Gallico," v. 12). The meaning of this is uncertain, and has been the fubject of much comment. We take it to mean that, befides the ordinary trees of France, there grew beech and Scotch fir as well in Britain. Geologically connected as our ifland had been with Holland and the neighbouring countries, it is inconceivable that the beech fhould not have been an indigenous tree, as it certainly is at prefent in Bucks, con-

ſidering the vaſt woods of beech which now wave
in Denmark. Our modern Scotch firs were undoubtedly introduced into England from Scotland
in the reign of James I.; .but many facts lead to
the concluſion that the tree had exiſted in early
times in our country and then become extinct.
Another interpretation regards the paſſage as
meaning that Britiſh vegetation was ſimilar to
that of France, except in poſſeſſing the beech and
fir. At any rate, the ſmall-leaved elm is of
Roman introduction, though the wych-elm is as
diſtinctively indigenous as are our two ſpecies
of Britiſh oak. The cypreſs and Oriental plane,
laurel and myrtle, ilex, ſumach and arbutus, are all
of them gifts from Rome. The true rhododendron
or oleander (*nerium* of Pliny) alſo came to us
from Rome, as the Romans had themſelves
received it from Greece. It is poiſonous, ſays the
old natural hiſtorian, to goats and ſheep, but a
remedy to man againſt the bites of ſerpents.

Many of our fruit-trees have an Eaſtern origin,
and came to us through Rome. The walnut
(*Anglicè*, "ſtrange nut") is an Eaſtern tree.
Thoſe of Perſia were regarded by the Romans as
being beſt, and were called "royal."[1] Like the
rice at preſent in faſhion at our marriages,
the walnut was highly eſteemed at nuptial cere-
monies of the ancients. The peach was alſo a
Perſian tree, hence called *perſica* by the Romans,
the French *pêche*, our peach. The apricot alſo
by its name teſtifies to the ſunny lands of its

[1] Pliny, "Hiſt. Nat.," xv. 22.

original home. Filberts have become fo common
in our copfes that they might be regarded as indigenous did we not know that the nut originally
came from Pontus to Rome, and was thus fometimes called the *nux Pontica.* Thofe of Thafos
were celebrated. It abounded in the diſtrict round
Avellano in Campania, whence comes the botanical name of the hazel, *corylus Avellana,* Lin.[1]
Quinces, mulberries, cheftnuts, and plums are
more benefactions of Rome, and of courfe vines
and figs. There were no cherries in Italy before
the victories of Lucullus. He brought them to
Rome, and in 120 years the tree penetrated
beyond the fea, fays Pliny,[2] into Britain. Befides
the five fpecies of rofes which Bentham deems indigenous, the *R. Gallica,* Ayrſhire, and China
rofes are alfo due to Roman commerce. Of other
flowers, the lily, *crocus vernus,* tulip, lilac, ranunculus, hyacinth, dianthus caryophyllus (clove-
pink), fweet-william, came from Rome.[3] Flax
and hemp, thofe moft ufeful allies of civilization,
came from the Eaft to Rome and thence to us.

Paffing from the flower to the kitchen-garden,
our peas and cucumbers came acrofs with the
Roman conquerors. Who does not remember
Virgil's Corycian old man, and his well-ordered
garden? and how

> " Tortus per herbam
> Crefceret in ventrem cucumis."[4]

---

[1] Pliny ; and Daubeny, " Trees and Shrubs of the Ancients,"
p. 6.
[2] Pliny, " Hiſt. Nat.," xv. 25.
[3] See Victor Hehn.   [4] " Georg.," iv. 121.

Leeks, onions, and garlic reached England also from Rome, as they came there from the East. In Egypt they were esteemed sacred, and even gods, so that oaths were taken upon them. Hence Juvenal lashes the Egyptians (xv. 9):

"Porrum et cæpe nefas violare ac frangere morsu;
O sanctas gentes, quibus hæc nascuntur in hortis
Numina."

It will be remembered how the Jews on leaving Egypt grumbled at missing "the fish, which we did eat in Egypt freely; the cucumbers, and the melons, and the leeks, and the onions, and the garlic" (Numbers xi. 5). Their fondness for these latter dainties gave them that "fœtor Judaicus" which was popularly ascribed to them by the ancients, and which the garlic-eating natives of Italy and Spain have now inherited. When Marcus Aurelius was travelling through Palestine into Egypt, he was much disgusted at the crowds of strongly-smelling Jews which flocked around him, and is said to have exclaimed: "O Marcomanni, O Quadi, O Sarmatæ, tandem alios vobis inertiores inveni!"[1] Mr. Darwin in his last book shews that the earthworm's greatest vegetable dainty is an onion. Herodotus saw engraved on one of the pyramids the exact amount which had been expended during its building on radishes, onions, and garlic for the workmen, and was told by his interpreter that the sum was 1,600 talents of silver.[2] The strip of land

[1] Amm. Marcel. xxii. 5, 5 (quoted by Victor Hehn).
[2] Herod., ii. 125.

bordering the Nile on each fide of its courfe which forms Egypt is nothing but a natural garden of the greateft fertility ; hence its ftore of grateful and appetizing vegetables.

The art of grafting and of confining fermenting liquors in jars with corks alfo came to Britain from Rome. Whether beer, the national drink of Wales and England, was firft brewed here by the Romans admits of a doubt. In Egypt, where there were no vines, the natives drank a wine made from barley, to which Æfchylus feems to allude when he makes his King of the Argives fay : " You will not find the men of this country drinking wine made of barley."[1] But it is more than probable, judging from what we know of favage races in Africa, that the Britons ftruck out the procefs of fermentation, whereby a certain kind of beer was produced, for themfelves. Virgil's words imply this in his beautiful picture of life in the North of Europe, which is yet true there in many points of Chriftmas revellings :

> " Advolvere focis ulmos, ignique dedere,
> Hic noctem ludo ducunt, et pocula læti
> Fermento atque acidis imitantur vitea forbis."

Although a popular couplet afcribes the coming of hops to England to a much later date, probably they, too, were firft imported by the Romans. The art of making butter has alfo been fometimes attributed to Rome ; but from the analogy of the Scythians and other paftoral

[1] Herod., ii. 77 ; Æfch., "Supp.," 953 ; Virg., "Georg.," iii. 376 (Victor Hehn).

nations, it is moſt likely to have been practiſed in England long before the Roman invaſion.

Ordinary wayſide plants and weeds, again, have been largely reinforced by recruits from Rome. Here, again, it is impoſſible clearly to ſettle which plants entered England with the Romans, and which came in after-days along with monkiſh contributions to Britiſh gardens; but one ſpecies (if not three) of nettles is certainly to be attributed to the earlier gardeners. Our red poppies, too, and a vaſt number of cornfield ſeeds, ſeem to have immigrated from the Italian farms. That rare Engliſh plant, the *aſtrantia major*, has been aſſerted to be another Roman immigrant. It is only found at preſent about Ludlow and Malvern. Laſtly may be added to the long liſt of Italian benefactions the knowledge of keeping bees in hives. Wild man everywhere feeds on honey, but to preſerve the ſtock near habitations, and at ſtated times to procure the produce of the ſwarms, is the teaching of civilization. Inſtruction in bee-keeping was a fitting gift from the nation which has produced the beſt poem on bees as yet known to the world.

There are no alluſions, either in Homer or the Bible, to the invention of hives. Meſſrs. Greenwell and Rolleſton[1] " learn from Profeſſor Weſtwood that, according to Spinola, our domeſtic ſpecies *apis mellifica* rarely occurs in Liguria ; and he ſuggeſts that this ſhews either that the Ligures

---

[1] "Britiſh Barrows" (Appendix on the Flora and Fauna of the Neolithic Period), Oxford, 1877, p. 719 *ſeq.*

were not the colonizers of Wales, as has been affirmed, or that they did not bring their bee *a. liguſtica* with them."

A few more notes may be added from the ſame authors' excellent Appendix, explanatory of ſome of our ſtatements. The aſh and beech are not indigenous in Scotland, though common now in ſome of the northern diſtriċts. They controvert Daubeny's ſtatement that the beech was not known in England until the Norman Conqueſt, and conſider that by the tree mentioned by Cæſar as *abies*, he meant the "ſilver fir." We regard it, however, as meaning the Scotch fir. The yew and the juniper were for ages the only other repreſentatives of the Coniferæ in the iſland. The ſmall-leaved lime they conſider as probably indigenous, if not the *tilia Europæa*. It was uſeful for matting, which is an invention older than weaving. In the "Romaunt of the Roſe," "The Aſſembly of Foules," and "The Complaint of the Blacke Knight," Chaucer gives three liſts of trees which may be taken as the repreſentatives of Engliſh woodlands in the fourteenth century. We will name ſome of theſe: laurels, pines, cedars, olives, pomegranates, nutmegs, almonds, figs, dates, in a "gardin" which ſeems a fanciful aſſemblage ; for he adds "many homely trees," peches, coines (quinces), apples, medlers, plommis, peris, cheſteinis, cheriſe, nottes, aleis (alise, Fr., the lote-tree), bolas (bullace), maplis, aſhe, oke, aſpe, planis, ewe, popler, lindis (limes), boxe, cypres, and

"the freſhe hauthorne
In white motley that ſo ſote doeth yſmell."

It is eafy from this lift to fancy the park and garden fcenery of Chaucer's times.

With regard to the common fowl, there is no evidence, they fay, of it in neolithic interments in this country. They hold that it probably came with the peacock by the way of Babylon to Greece and Rome rather than by the Red Sea. It is known from Buddhift writings that the ancient Indian merchants took peacocks to Babylon. As a pendant to Chaucer's vegetation the following pre-Roman landfcape may be cited: "The contemplation of a herd of dark-coloured mountain cattle in the north of this country, of fmall fize, and yet with ragged 'ill-filled' out contours, ftanding on a wintry day in a landfcape filled with birch, oak, alder, heath and bracken, has often ftruck me as giving a picture which I might take as being very probably not wholly unlike that which the eyes of the ancient Britifh herdf-man were, familiar with " (p. 744).

CHAPTER XII.

IT is not furprifing that Virgil, with his keen fenfe of natural beauty and exquifite play of fancy, loved to lighten his verfe with the fwift wings and happy fongs of birds. All poets turn naturally to thefe artlefs fongfters. But Virgil's lines betray here and there that he loved and ftudied the ways of his native birds in a manner very unufual in his time. Birds are introduced, indeed, in his pictures of country life, or as illuftrations of human pathos in the conventional manner of ordinary poets, as he had inherited the cuftom from Homer, and as Pope did in the laft century ;[1] but the felicitous images and wording of many paffages fhew that he had clofely ftudied bird-life, and feized upon new and ftriking traits in it for the embellifhment of his poems. Born at Andes, now Pietola, a hamlet near Mantua,

---

[1] See Pope's celebrated lines, for inftance, on a pheafant in the "Windfor Foreft," and the "lonely woodcock," "clamorous lapwing," and " mounting lark " of the fame paftoral.

and ſpending much of his manhood as well as all
his impreſſionable youth in that diſtrict of marſh
and hill, while the Po, Father of waters, rolled
along but a few miles from his father's houſe ;
with glimpſes of diſtant mountains, now ſhrouded
in miſts, now painted with the flying tints of
morn and evening, while murmurs of pines and
running waters were everywhere around him, it
was only natural that the birds which haunted his
native fields ſhould become dear to Virgil. In ſpite
of more than one attempt of the lawleſs ſoldiers,
whom the chances of civil war had planted in his
neighbourhood, to diſpoſſeſs him, he ſeems to have
dwelt, more or leſs, on his father's eſtate till he died.
In ſooth, it was a fit home for a poet, a realization
of an Engliſh brother's dream centuries afterwards:

> " Happy the man whoſe wiſh and care
> A few paternal acres bound,
> Content to breathe his native air
> In his own ground.

> " Sleep found by night ; ſtudy and eaſe
> Together mix'd ; ſweet recreation,
> And innocence, which moſt does pleaſe
> With meditation."

The ancients, aided by hints from his own
writings, depict for us Virgil's home as having been
about three miles from the city Mantua, on high
ground, running down towards it ; and the eſtate—
that "angle of ground which had charms beyond
all others for him"—as ſpreading over the roots
of the hilly diſtrict between the Mincio and Po;
on the upper part ſcant of herbage and ſtony, on
the lower ſomewhat marſhy and low-lying :

" Qua fe fubducere colles
Incipiunt, moHique jugum demittere clivo,
Ufque ad aquam et veteres jam fraɗa cacumina, fagos."[1]

Thus it combined for the great Latin poet thofe
ftriking features of mountain and marfhland which
in our own days have refpeɗively nurtured a
Wordfworth and a Tennyfon. Many paffages
from Virgil's poetry could be pointed out in which
he has felicitoufly depiɗed the fcenery of both
diftriɗs; and in thofe of the marfh country fome
touches remind us at times of a kindred art, as
feen in Mr. Millais's beautiful piɗure of "Chill
Oɗober." For inftance ("Ecl.," i. 48):

" Quamvis lapis omnia nudus,
Limofoque palus obducat pafcua junco ;"

and (*ibid.*, 56) the boundary hedge:

" Vicino ab limite fepes
Hyblæis apibus florem depafta faliɗi ;"

and more clofely ftill ("Ecl.," vii. 12):

" Hic virides tenera prætexit arundine ripas
Mincius, eque facra refonant examina quercu."

Among all his defcriptions of bird-life it is fmall
wonder that he loved moft ardently the birds of
marfh and river-fide, fwans, cranes, halcyons, and
the like on the one hand; and thofe of the cliff
and bare hillfide on the other, eagles and hawks;
but the former clafs decidedly predominates. As
for the fmaller tribes of twittering fongfters in-
habiting the ordinary bufhes and brakes, he feems
not to have beftowed a thought on them. A

[1] "Ecl.," ix. 7.

bird to deſerve his notice, and win an endleſs life in his melodious verſe, muſt not only be of a marked and fine ſpecies, but alſo, in moſt caſes, one which has obtained fame from Homer and other poets and writers dear to Virgil. If our own Milton had one favourite bird, the nightingale, whoſe praiſes he ſings with an iteration as beautiful as the ſongſter's own ſtrains, Virgil's favourite was unqueſtionably the eagle. But his eagle does not ſit tamely by the throne of Jupiter while the queen of heaven careſſes its neck; it is eſſentially a bird of daring and rapine and ſolitude. Like Tennyſon's eagle—

> " He claſps the crag with hooked hands,
>   Cloſe to the ſun in lonely lands,
>   Ring'd with the azure world he ſtands ;"

or like Shelley's eagle, "ſoaring and ſcreaming round her empty neſt," ſhe

> " Could ſcale
> Heaven, and could nouriſh in the ſun's domain
> Her mighty youth with morning ;"

or it reſembles, in another mood, Mrs. Browning's

> " Eagle with both grappling feet ſtill hot
> From Zeus 's thunder."

All the power and ruſh of Virgil's fineſt verſe is ſpent in picturing the eagle to his hearers, as he muſt have often ſeen it ſweeping down from the ſpurs of the Alps round Lago di Garda, and carrying off its hapleſs victim, ſwan or marſhſnake, over the wide valley to the diſtant creſts of the Apennines. Perhaps the very vigour and sublimity of the king of birds endeared it to him who must have been conscious that, however musical and polished were his verses, they seldom soared into the empyrean of poetry.

If the influences of the scenery amidst which his home-life was spent were twofold, so the landscapes and the figures with which he peoples them in his poems partake also of a double character. They are at once natural and conventional; natural so far as they reflected the low-lying pastoral country in the basin of the Po; conventional when coloured with reminiscences of Theocritus, and planted in a Sicilian *entourage*. Besides these characteristics of his verse, it is frequently set with fanciful or "otiose" epithets and animals. Thus lynxes, lions, and lionesses, wild asses, scaly dragons, painted birds, and the like, frequently adorn its flow. Over and above this poetical surplusage, however, the student of Nature will detect much close observation, especially of birds, in Virgil's lines. Like his own Helenus ("Æneid," iii. 361), "he knows the voices of the birds and the omens to be derived from their swift flight," and we shall pause before accusing him in any of his delineations of bird-life as drawing only upon his imagination, or adding merely conventional touches, lest our fancied wisdom should incur the charge of foolish censoriousness which Aulus Gellius brings against one Higinus, who ventured rashly to criticize Virgil's ornithology.[1] Wider reading, and more careful study, will, on the contrary, point out more

[1] Aulus Gell., vi. 6, 5.

beauties in Virgil's ornithological pictures. The exact manner in which he defcribes, often in a line, the chief characters of a bird, and adds a new delight to its traits by fome play of fancy— fome lively touch of imagination—becomes very apparent on a furvey of his poems. Scientific ornithology, of courfe, no one would look for in a poet, efpecially a poet of Virgil's age. Has the " fea-blue bird of March " been ever fatisfac- torily identified in the Laureate's poetry ? Virgil alludes to the migration of birds, for inftance, once or twice, but never troubles himfelf to enunciate a theory upon their departure or return. They bring back fpring on their wings, and return to their fweet nefts and dear offspring, and that is enough for him. In a fimilar manner we fhall content ourfelves with pointing out the nice ob- fervation and the poetic mind with which fome of the birds of North Italy are defcribed in his verfe. The furvey will fhew how eminently naturaliftic is his poetry in the midft of fo much that is imitative and conventional.

It may be faid generally that the "Eclogues" and "Georgics" exhibit a more genial fancy, and more ftriking images of bird-life, but that the "Æneid," as befits a work of mature years, is ftudded with more carefully finifhed workmanfhip. That the poet was continually improving, and adding frefh touches to it, is proved by his folicitude concerning the poem at his death, and his wifh that it fhould be burnt after his deceafe, as not fatisfying his own ideal. With thefe preliminary remarks, we fhall now difplay to lovers of a poet dear to every cultivated mind, the contents of the Virgilian aviary.

To begin with the birds of the lowland and marfh, our own carrion crow, which fo often reforts to the edges of rivers and the feafide for fhellfifh and muffels, had frequently brought the poet good luck by cawing from fome hollow oak on his left; or, wicked thief that it was, called for rain, with full clear voice, as it ftalked along fingly on the dry fandbanks.[1] This bird poffeffed a great reputation amongft the Romans for prophetic and thievifh powers. From its ufual cuftom of attacking its prey firft in the eyes, came a Latin proverb, "To dig out the eyes of crows," anfwering to ours about catching a weafel afleep. It was celebrated, too, for living long, fharing this fame with the ftag, and eagle, and the ferpent which could put off years with its fkin. Its eyes were ufed by the profligate as a love-charm to throw duft in the eyes of hufbands.[2] Its larger relative, the raven, was alfo fuppofed to have an inftinctive knowledge of the approach of fine weather:

> "Thus thrice the ravens rend the liquid air,
> And croaking notes proclaim the fettled fair.
> Then, round their airy palaces they fly
> To greet the fun, and feized with fecret joy,

---

[1] "Ecl.," ix. 14; "Georg.," i. 388. Cnf. Hor. "Od.," iii. 12:

> "Aquæ nifi fallit augur
> Annofa cornix."

[2] Prop., iv. 5, 15.

When ftorms are over-blown, with food repair
To their forfaken nefts and callow care.
Not that I think their breafts with heavenly fouls
Infpired, as man, who deftiny controls,
But with the changeful temper of the fkies,
As rains condenfe and funfhine ratifies,
So turn the fpecies in their altered minds
Compofed by calms and difcompofed by winds.
From hence proceeds the birds' harmonious voice,
From hence the crows exult and frifking lambs rejoice."[1]

Atmofpheric changes connect themfelves, in Virgil's
mind, with the changed behaviour of birds. So,
when wind is impending:

"Back from mid ocean home the cormorants fly
With clamours, and the coots where fands are dry
Refort, while herons love the upper fky."[2]

Or when rain is imminent:

"Huge flocks of rifing rooks forfake their food
And, crying, feek the fhelter of the wood.
Befides, the feveral forts of watery fowls
That fwim the feas or haunt the ftanding pools,
Then lave their backs with fprinkling dews in vain,
And ftem the ftream to meet the promifed rain."[3]

Cranes view it blowing up, and defcend from
their lofty flights to the deep valleys with much
noife. And elfewhere he compares the buftle
infide a beleaguered city to their fcreaming:

"Juft fo 'neath inky clouds
Strymonian cranes fcream, cleaving lofty fkies
With clamour, 'fcaping rain with joyous notes."[4]

The notion comes originally from Ariftotle, who
fays that cranes fly at a great height, in order that
they may difcern things far off; and if they fore-

---

[1] Dryden, "Georg.," i. 410.   [2] *Ibid.,* i. 361.
[3] *Ibid.,* i. 381.
[4] "Georg.," i. 374 ; "Æn.," x. 264.

fee ſtorms and wintry weather, they deſcend and
reſt on the ground. Akin to the cranes is the
ſtork, and in ſpring "the white bird comes which
is hated by long ſnakes."[1]   It is indeed difficult
for the dweller by Mincius, "clothed in glaucous
reeds," to forget the birds of the river—

"Around, above,
Birds of the bank or river-bed in plumes
Of party-coloured ſplendour ſoothe the ſkies
With ſong, and flit by ſtream or woodland lawn."[2]

The wild-gooſe had probably proved deſtructive
to the poet's crops, for he terms it "improbus
anſer" (which the late Dr. Sewell quaintly tranſlates
"the caitiff gooſe"), and ſcoffs at its attempts at
ſinging amongſt ſwans.   The wild ſwan, with its
graceful form and not unmuſical notes, is, on the
contrary, a ſpecial favourite with Virgil.   Here is
a ſtudy of wild ſwans flying home:

"Like a long team of ſnowy ſwans on high
Which clap their wings and cleave the liquid ſky,
While homeward from their watery paſtures borne,
They ſing and Aſia's lakes their notes return.
Not one who heard their muſic from afar
Would think theſe troops an army trained to war,
But flocks of fowl that when the tempeſts roar
With their hoarſe gabbling ſeek the ſilent ſhore."[3]

Although Dryden was an accompliſhed fiſherman, his rendering of the above lines proves him
to have been no ornithologiſt.   He ſucceeds
better in relating the transformation of Cycnus
into a ſwan:

"Love was the fault of his famed anceſtry,
Whoſe forms and fortunes in his enſigns fly.

[1] "Georg.," ii. 320.   [2] "Æn.," vii. 32.
[3] Dryden, "Æn.," vii. 699.

For Cycnus loved unhappy Phaeton
And ſung his loſs in poplar groves alone,
Beneath the ſiſter ſhades to ſoothe his grief
Heaven heard his ſong and haſtened his relief ;
And changed to ſnowy plumes his hoary hair,
And winged his flight to chant aloft in air."[1]

More than one of Virgil's ſimiles of ſwans attacked by eagles may have been in the mind of Sir E. Landſeer, when he painted his picture of this ſubject, which ſome fifteen years ago was the ornament of the Royal Academy.

"So, twice ſix ſwans in line exulting ſee,
Whom Jove's bird ſwooping from the upper ſkies
Has ſcattered, now the band or gains kind earth,
Or looks down on it as though gained."[2]

And again :

"As when Jove's thunderbearer's crooked claws
Seizing on hare, or ſwan with whiteſt breaſt,
Bears it aloft."[3]

And once more :

"Bathed in red evening ſkies, Jove's tawny bird
Was hunting ſhore-birds and the clanging crowd
Of hurrying ſwans, when ſudden downward ſhot
He ſmites a goodly ſwan into the waves
And bears it off, bold thief with crooked legs."[4]

And he ſpecially ſpeaks of the plain near Mantua :

"Where feed the ſnow-white ſwans on graſſy ſlopes."[5]

Another water-bird is introduced in the " Æneid," iv. 253, which at firſt ſight, from its ſplaſhing dive, might reſemble the oſprey, of which a few ſpecimens may yet be ſeen in Roſſ-ſhire ; but the

[1] Dryden, " Æn.," x. 189.
[2] *Ibid.*, i. 392.
[3] *Ibid.*, ix. 562.
[4] *Ibid.*, xii. 247.
[5] " Georg.," ii. 199.

word *humilis* probably points to the ſtraight, low-flying advance of a cormorant over the waters. Mercury is depiĉted as plunging into the ſea, juſt as Homer had ſung in the "Odyſſey" (v. 57):

> "Headlong the god dived quick into the waves,
> Like the low-flying bird which round the ſhores
> And round fiſh-haunted rocks flies near the ſea."

The poet had certainly obſerved with care the haunts of the cormorant, and in another paſſage accurately draws them ("Æneid," v. 128):

> "Far out at ſea againſt the foam-white cliffs ·
> Glooms a dark rock oft ſmit by ſwelling waves,
> When winter's ſtorm-winds blind the ſtars ; but raiſed
> In calm-flowing ſeas above their level tides,
> It forms a ſtation much of cormorants loved,
> Where grateful ſunſhine laves them."

Pigeons, again, are birds for which Virgil had a ſpecial liking. He ſpeaks of the Chaonian pigeons fluttered at the approach of an eagle. And his Damon ſays :

> "To the dear miſtreſs of my love-ſick mind,
> Her ſwain a pretty preſent has deſigned ;
> I ſaw two ſtock-doves billing, and ere long
> Will take the neſt, and hers ſhall be the young."[1]

And again :

> "Stock-doves and turtles tell their amorous pain,
> And from the lofty elms of love complain."[2]

Though the reader of the original ſcarcely recogniſes this for the tranſlation of words ſo true to Nature as, "Not in the meantime ſhall the wood-pigeons, ſo dear to thee, hoarſe with cooing, and the turtle, ceaſe to moan from their lofty

[1] Dryden, "Ecl.," iii. 69.   [2] *Ibid.*, i. 58.

elm." Another beautiful image deſcribes Hecuba and her daughters flying to the altars, when Troy was taken, like pigeons flying wildly from the black ſtorm ("Æneid," ii. 516). But perhaps his fineſt ſtudy of the pigeon deſcribes the rock-dove darting from her cave, as we may obſerve it on our own cliffs at Speeton or Cromarty:

> " As, ſudden ſtartled from her cave, the dove
> Whoſe dear abode the darkling pumice hides,
> Cleaves the air ſwiftly, flapping through the cave
> Till all its roof reſounds, but ſoon, borne on,
> Lightly ſkims o'er the liquid plain, nor moves
> Her pinions fleet."[1]

This is felicitouſly true to Nature. Eye and ear are alike ſatisfied, and it ſeems to bring the ruſh of air and roar of waves round the baſe of the ſea-cliffs to the mind as it is read. Another ſimile relates what too frequently befalls ſuch a bird on its emerging from the cavern's gloom, and is another highly finiſhed picture:

> " With equal eaſe the ſacred hawk purſues,
> And ſweeping upwards from his naked crag,
> High o'er a flying cloud ſtrikes down the dove,
> Then grips and tears her with his crooked claws
> Till gore and feathers float off down the breeze."[2]

A ſimilar reminiſcence ſtrikes the poet as he thinks of Tarchon triumphantly bearing off booty:

> " So, high aloft the tawny eagle ſweeps,
> Bearing away the ſerpent ſhe has ſeized,
> Wraps her feet round it and drives in her claws.
> Wounded but dauntleſs ſtill the angry ſnake
> Twines his thick folds and briſtling with ſet ſcales,
> Hiſſes and rears his threat'ning creſt ; but ſhe
> Continues ſtriking with her crooked beak,
> O'erwhelms his rage, and wings the ſounding air."[3]

---

[1] " Æn.," v. 213. [2] *Ibid.*, xi. 721. [3] *Ibid.*, xi. 751.

Compare, too, the beautiful lines in "Æneid," xi. 721, *feq.*

Among water-birds, Virgil does not dwell much upon the halcyon, though it poffeffed what we might fancy fo attractive a fet of myths. In a picture of a fummer evening, he makes the fhores refound with the halcyon, the brakes with the goldfinch, and tells how, in the beginning of fine weather, the halcyons, beloved by Thetis, fpread their wings on the fhore to the warm fun" ("Georg.," iii. 338; i. 398). He has beautifully touched the fad tale of the nightingale in two paffages, relating in the firft how Philomela, after ferving her dreadful banquet to Tereus, fled to the wildernefs on the very wings with which fhe had fluttered in her mifery round home; and in the fecond, comparing the fad ftrains of Orpheus, bereft of his wife, to the lorn nightingale, with a happy imitation of the tendernefs of the celebrated paffage in the "Odyffey":

> "As the lone bird of fong in poplar fhades
> Bewails her ravifhed young, which fome hard clown
> Noting hath drawn, ftill fledglings, from their neft;
> So fhe weeps night-long, and from fome thick bough
> Again renews her ftrain, her ftrain fo fad,
> And fills wide filence with her forrowing plaints."[1]

Progne, Philomela's fifter, as well from the myth as from being the familiar bird of houfe and lake, is not forgotten. She is among the birds harmful to bees, "bee-eaters and other birds and Progne" (*i.e.* the chimney-fwallow), "marked on her breaft by bloody hands" ("Georg.," iv. 14).

[1] "Ecl.," vi. 80; "Georg.," iv. 511.

Again, "with fhrill cries fhe flits around the lakes" (Georg.," i. 377), "and hangs, with many a twitter, her neft on the rafters" (*ibid.*, iv. 307). But a ftill more famous paffage occurs in the "Æneid," xii. 473, concerning which Gilbert White writes pleafantly, but as a practifed naturalift, in his "Selborne" (ed. Bell, vol. i. 166). After remarking that the ancients were not wont to difcriminate between different fpecies as we are, he concludes from many little touches in the picture, that the poet (as in the two inftances quoted already), was referring to the chimney-fwallow rather than to its, comparatively fpeaking, more clumfy brother, the martin:

> "As when the dufky fwallow darts athwart
> Some rich man's fpacious halls and lofty courts
> To catch on nimble wings her tiny prey,
> Then bears it fpeedy to her prattling neft,
> And now by empty portico fhe gleams,
> Now twitters by the low-lying marfh."

The woodpecker (*picus*) is happily connected with another myth. Dryden's poetry is, again, better here than his ornithology:

> "Circe long had loved the youth in vain,
> Till love refufed, converted to difdain ;
> Then, mixing powerful herbs, with magic art
> She changed his form who could not change his heart,
> Conftrained him in a bird and made him fly
> With parti-coloured plumes, a chattering pie."[1]

The owl is another Virgilian bird. There are at leaft four fpecies of fmall owls in Italy ; but the poet generalizes them in the few yet telling lines

[1] Dryden, "Æn.," vii. 189.

which he devotes to them.   When fine weather is
imminent :

> "In vain from fome high roof the mournful owl,
> Watching the funfet, hoots till night grows late ;"

and,

> "Lone on the roof with deathful cries the owl
> Oft wails, prolonging with fad moans her grief ;"

and once more,

> "On tombs at times and ruined gables late,
> Wailing to darknefs, fits th' ill-omened bird."[1]

A ftriking paffage in the firft "Georgic," 404, is
another fign of Virgil's fondnefs in his poetry for
affociating birds with popular myths.   It relates
to the ofprey, or more probably fome kind of
falcon, purfuing Ciris—another unknown bird.
The ftory of Nifus and his daughter Scylla is
told in Ovid, and may be found in the "Ciris,"
elaborated from Virgil's own few lines in this
paffage :

> "Towering aloft avenging Nifus flies,
> While dared below the guilty Scylla lies.
> Wherever frightened Scylla flies away,
> Swift Nifus follows and purfues his prey.
> Where injured Nifus takes his airy courfe,
> Thence trembling Scylla flies and fhuns his force.
> This punifhment purfues th' unhappy maid,
> And thus the purple hair is dearly paid."[2]

It may be worth while to fay that the word
"dared" in the fecond line of this tranflation is a
technical term of hawking; meaning that a bird
lies clofe to the ground in terror at fome enemy
foaring above it.

---

[1] "Georg.," i. 403 ; "Æn.," i. 404 ; xii. 862.
[2] Dryden.

This concludes the liſt of birds which were dear to Virgil. A few more lines relate to their economy, their uſe in augury, and the like. Thus a pretty picture gives us the woodman felling ancient trees, and deſtroying in their fall the time-honoured neſts of birds; and another, the lonely thickets enlivened in ſpring with their ſong. Occaſionally ſome virulent diſeaſe attacks them, and then "the very air is inhoſpitable, headlong in death they drop from the lofty clouds;" or winter's ſtorm, and the approach of night drifting downwards from the mountains, drives them in thouſands to take ſhelter in their leafy coverts; while at times theſe troops of birds (perhaps ſtarlings were in Virgil's mind), ſettle down on the thick plantations, and hoarſe flocks of ſwans, in the noiſy ſwamps of fiſhy Po, make the ſky reſound with their cries ("Æneid," xi. 456, etc.). In order to adorn the lowly home of Evander ("Æneid," viii. 456), a touch is added which nearly approaches the poetic feeling of modern times; "the morning ſongs of early birds beneath his roof-tree" awake him. The fineneſs of Virgil's genius, the poetic colouring which he gives to all that he touches, are very apparent in theſe ſtudies of his birds. It is very true, indeed, that moſt of his ſimiles are drawn from Homer; but how often does he lend them a graceful turn which is wanting in the rough vigour of the original! "Take from Virgil," ſays Coleridge in the "Table Talk," "his melody and diction, and what is there of him?" A novel and enlarged method of obferving Nature, and the difcovery of a new fource of adornment for poetry, are at all events features peculiar to him. Modern ornithologifts owe to him, as has been fhewn, not a little; and all lovers of the country love it the better as they affociate its birds of paftoral fcenes with the mufical verfe and clear poetic infight of the great Roman poet.

CHAPTER XIII.

ROSES.

"An tu me in viola putabas aut in rofa dicere ?"
(Cic. *Tufc.*, v. 26.)

SO ancient and widely prevalent are the notions connected with the word "rofe," that it might well be queftioned whether "the rofe by any other name would fmell as fweet." The name comes to us, with flight dialectical variations, through Latin and Greek from the Arabic. Not that the Eaft is the exclufive home of the flower, for it is found in almoft every country of the Old and New World—from Norway to the North of Africa, and from Kamfchatka to Bengal. There are no rofes, however, in South America or Auftralia; but the greateft beauty and moft luxuriant growth of this lovely flower are un-doubtedly to be feen in the Eaft.

"Who has not heard of the Vale of Cafhmere,
    With its rofes the brighteft that earth ever gave ?"

All through the Bengal Prefidency rofes are magnificent; but their beauty culminates at Umritzur, which is a mafs of myrtles and rofes, like a city of the "Arabian Nights."[1] Of the many natural varieties, three are mainly the parents of the enormous number of kinds cultivated by modern gardeners, and thefe three were probably equally well-known to the ancients. Thefe are *Rofa centifolia*, which has been found wild in thickets on the eaftern fide of the Caucafus; *R. Damafcena*, a native of Syria; and *R. Indica*, the Chinefe rofe. Some 3,000 fpecies are now in cultivation in France, which will give an idea of the varieties which have fprung from budding, grafting, and feed; and Mr. Rivers enthufiaftically anticipates, it may be ftated for all lovers of the queen of flowers, that "the day will come when all our rofes, even mofs-rofes, will have evergreen foliage, brilliant and fragrant flowers, and the habit of blooming from June till November."[2] The rofe twice mentioned in the Old Teftament is no true rofe, but moft probably the narciffus. Similarly the fo-called Rofe of Jericho (*Anaftatica Hierochuntina*) is a cruciferous plant, found in

[1] Together with Adrianople thefe two cities make moft of the Oriental attar of rofes. Umritzur "makes attar of rofes from the *R. centifolia*, which only bloffoms once a year, and it makes it for the world. Ten tons of rofe-petals are ufed annually in it, and are worth from £20 to £30 per ton in the raw ftate. The petals are diftilled through a hollow bamboo into a veffel which contains fandal-wood oil. The contents are then poured out and allowed to ftand till the attar rifes to the furface in fmall globules, and is fkimmed off. The pure attar fells for its weight in filver."—"Greater Britain," i., p. 278.

[2] See Darwin, "Animals and Plants, etc.," vol. i., p. 391.

ſandy ſoil in Egypt and Paleſtine, juſt as our own
Chriſtmas-roſe is really the black hellebore.

The Romans by no means attached to their
gardens the ſenſe of a leiſurely retreat, full of
beautiful flowers and ſhade, as we do. The Latin
word for a garden, *hortus* (which is but a ſoftened
form of χόρτος), ſhews that they regarded it
mainly as a place for growing food; in ſhort,
their garden was orchard, kitchen-garden, and, to
a very ſmall extent, flower-garden in one.[1] This
economical view of a garden was a natural out-
growth of the practical Roman mind, although it
is ſeen, albeit in a minor degree, in the Greek
character as well. Roman gardeners, however,
rejoiced in beds of violets and roſes as much as we
do. Roſes were even forced in greenhouſes, ſo
that lovers of flowers might have them during
winter.[2]

> " Dat feſtinatas, Cæſar, tibi bruma coronas ;
>   Quondam veris erat nunc tua facta roſa eſt."[3]

> " Once, Cæſar, ſpring was wont thy flow'r to greet ;
>   Now winter's roſes hurry thee to meet."

Beſides miniſtering to the pleaſures of a garden,
roſes were largely uſed at Rome for garlands, to

---

[1] Cnf. Cicero, "Cato Major," caps. xv., xvi., where with
many expreſſions which ſpeak of the delight in ſunſhine and
ſhade of the country, the key-note is ſtruck by the words, " Jam
hortum ipſi agricolæ ſuccidiam alteram appellant."

[2] Compare Cicero, "Cum roſam viderat, tum incipere ver
arbitrabatur " ("Verr.," ii. 5, 10) ; and the philoſopher Seneca's
indignant queſtion, "Non vivunt contra naturam qui hieme
concupiſcunt roſam ?" (Ep. cxxii. 8.)

[3] Mart., xiii. 127. See Becker's " Gallus," p. 289, ed. 1844.

be worn during the caroufals which followed the
chief meal of the day.   As early as the fecond
Punic war this feftive cuftom prevailed.   There
was a notion among the Greeks that the flowers
prevented intoxication; but they were chiefly
fubfervient to luxury.   Befides rofes, violets were
alfo ufed for garlands, together with the green
leaves of the myrtle, ivy, and parfley.   It was
ufual for the hoft to fupply thefe garlands, much
as a modern entertainer places a fmall nofegay
before each of his guefts.   Everyone will remember
the beautiful little ode of Horace, in which he
warns his fervant againft extravagance in the
matter of garlands, bidding him refrain from feek-
ing where "the laft rofe of fummer" delays; nor
has he written a more tender idyl than that which
fhews us Pyrrha binding up her golden hair, while
fome flender youth courts her in a grotto hung
with rofes.[1]   Indeed, the rofe has always been the
flower moft dear to poetry.   "Place a hundred
handfuls of fragrant herbs and flowers," fays the
Perfian Jami, "before the nightingale, yet he
wifhes not in his conftant heart for more than the
fweet breath of his beloved rofe."   It was a
favourite flower of Milton, owing to his claffical
reading.   In Eve's nuptial bower,

> "Each beauteous flower,
> Iris all hues, rofes, and jeffamine,
> Reared high their flourifh'd heads between, and wrought
> Mofaic."

---

[1] It is fcarcely neceffary to add that Milton has tranflated
this ode of Horace into as dainty Englifh as the original, and
in the fame metre.

There the firft parents,

> "Lulled by nightingales, embracing flept,
> And on their naked limbs the flowery roof
> Shower'd rofes which the morn repaired."

Eve is painted,

> "Veiled in a cloud of fragrance, where fhe ftood,
> Half fpied, fo thick the rofes blufhing round
> About her glow'd."

And when Adam firft learns his wife's tranf-greffion :

> "From his flack hand the garland wreath'd for Eve
> Down dropt, and all the faded rofes fhed."[1]

Shakefpeare's rofes are thofe which bloffomed on the hedges by the Avon, and in the little cottage-plots with which he was moft familiar. His "fweet mufk-rofes" are the wildings of his own country lanes. He has ftamped an indelible affociation on this flower by relating the ftory of red and white rofes becoming the badges of the rival houfes of York and Lancafter (" 1 Henry VI.," ii. 4). All who have read the beautiful "Virgin Martyr" of Maffinger will remember how felici-toufly he makes ufe of the legend which tells that rofes were fent down from Paradife to ftrengthen the martyr's refolution.

The rofe was efpecially facred to Venus. She was fabled to have rifen from the fea dropping rofes over Rhodes, itfelf named from and famous for that flower.[2] Another legend told that fhe

---

[1] "Par. Loft," iv. 697, 771, and ix. 425.
[2] Ovid, "Faft.," v. 354 :
> "Et monet ætatis fpecie, dum floreat, uti ;
> Contemni fpinam, cum cecidere rofæ."

prefented a rofe to the Egyptian God of Silence,
Harpocrates, whence the expreffion "under the
rofe."[1] It was ufed at Rome on all feftive or
folemn occafions, and is frequently alluded to by
the Roman poets in reference to its beauty and
the moral its frailnefs pointed, as, indeed, the
poets of every nation have fung. Thus Horace
fpeaks of the " nimium breves flores amænæ rofæ;"
and Martial, when addreffing his own book of
poems :

> " Hæc hora eft tua, cum furit Lyæus.
> Cum regnat rofa, cum madent capilli."[2]

The expreffions "to lie among rofes," to
" drink," or " live " among them, were fynonyms
at Rome for luxurious living; and Cicero thus
paints the exceffive luxury of Verres : " Lectica
octophoro ferebatur, in qua pulvinus erat perluci-
dus, Melitenfi rofa fartus; ipfe autem coronam
habebat unam in capite, alteram in collo, reti-
culumque ad nares fibi admovebat tenuiffimo lino,
minutis maculis, plenum rofæ" (Verr., ii. 5, 27).
" Rofa," or " mea rofa," became, naturally, a term
of endearment, juft as with us it has become a
Chriftian name. The annual dreffing of the
graves with flowers, which is fo well-known a
cuftom in modern France, fprang from the feaft of
rofes at Rome—the *rofalia*, or *rofales efcæ*, when
the tombs were adorned in like manner with

---

[1] Billerbeck, "Flora Claffica" (Leipzig, 1824), p. 132. " So
we condemn not the German cuftom, which over the table
defcribeth a rofe in the ceiling."—(Sir T. Browne, " Vulgar
Errors," v. 22.)

[2] Martial, x. 19, 19.

garlands of rofes. "Cato, in his 'Treatife of Gardens,' ordained as a neceffary point that they fhould be planted and enriched with fuch herbs as might bring forth flowers for coronets and garlands."[1] Pliny adds, however, that the Romans were acquainted with very few garden flowers for garlands fave violets and rofes. The *rofeta*, or rofe-beds, in which thefe rofes were grown, are much celebrated in Latin poetry, particularly thofe of Pæftum, which ftill delight the traveller,[2] and were renowned for bloffoming twice in the year.

Pliny is the chief authority for Roman rofes. He mentions that twelve varieties of the flower, all more or lefs efteemed, were known at Rome. Thofe grown at Prænefte and Capua were regarded as the beft. A botanical characteriftic of the rofe family is the poffeffion of five petals. Pliny had noticed this: "The feweft leaves that a rofe hath be five; and fo upward they grow ever ftill more and more, untill they come to thofe that have an hundred, namely about Campain in Italy, and neere to Philippos, a city in Greece, whereupon the rofe is called in Latine Centifolia." They have been brought to this fize, and to the fragrance which many of them, efpecially thofe of Cyrene, poffefs, he adds, "by many devifes and fophiftications" of the gardeners. Yet how little he knew practically about rofe-cultivation is apparent from his words, "the rofe-bufh loveth not to be planted

---

1 Pliny, "Nat. Hift." (Holland), xxi. 1.
2 "Biferi rofaria Pæfti," Virg., "Georg.," iv. 119 ; Prop., iv. 5, 59 ; and "punicea rofeta," Virg., "Ecl.," v. 17.

in a fat and rich foile, ne yet upon a vein of cley,"
which is the exact oppofite to the recommendations
of modern horticulture. Another hint may be
commended to the attention of rofarians: "They
that defire to have rofes blow betimes in the yeare
before their neighbours, ufe to make a trench
round about the root a foot deep, and poure hot-
water into it, even at the firft, when the bud of
the rofe beginneth to be knotted."

In fpeaking of the "wine rofat," or "oile
rofat," compounded of rofes, Pliny feems to mean
what we call attar of rofes, or rofe-water. The
beft rofe-water is at prefent made at Ghazeepore,
and it is ufed in much the fame manner as the
Romans employed their "wine rofat," for bathing
any fore or inflamed part of the body. But, as
ufual, Pliny recommends every part of the rofe
for different ailments. The root of a kind of
wild rofe (our dog-rofe, fo named from this fuperftition), is a fovereign remedy againft the bite of a
mad dog. "The afhes of rofes, burnt, ferve to
trim the haires of the eiebrowes. Dried rofe-leaves
do reprefs the flux of humours into the eies. The
flowre procureth fleepe. To rub the teeth with
the feed eafeth the toothach. The wild rofe-
leaves, reduced into a liniment with Beares greafe,
doth wonderfully make haire to grow again;"
thefe will ferve as fpecimens of the medicinal value
of the rofe in Roman eyes.[1] In Gerard's "Herbal"
will be found two folio pages of the medicinal
value of rofes in the eftimation of our forefathers.

---

[1] See "Nat. Hift.," viii. 41 ; xxi. 19.

That the rofe came from the Eaft to the Greeks, is teftified by the fact of Homer knowing nothing of the rofe as a flower. He did, indeed, know of attar of rofes, for ("Iliad," xxiii. 186) he makes Aphrodite anoint the corpfe of Hector with "oil of rofes."[1] In his time, the rofe itfelf had not been imported into Greece. The fame fact is evidently alluded to by his conftant ufe of "rofy-fingered" as an epithet of the dawn (which may be compared with our own poet's "God made Himfelf an awful rofe of dawn"), and of Aphrodite herfelf. Thus the introduction of her worfhip into Greece has been actually afcribed to the Phœnicians, who, we know, did bring there the planetary worfhip of the Affyrians. Moreover, "Aphrodite is placed by Homer in relation with the Charites, Eaftern perfonages, whofe name correfponds with the Sanfcrit Harits, meaning originally 'bright,' and afterwards the horfes of the dawn."[2] It is curious that the rofe, fave with the lyric poets, does not feem to have been a great favourite. Sophocles prefers the hyacinth. The dramatic poets, concentrating their thoughts on the tragedy of man's feeling and actions, difregarded it as a creature of a wholly different, a lower and a frivolous world. Anacreon naturally celebrates the flower, and does fo more than any other Greek finger:

---

[1] "Poeta rofam non norit, oleum ex rofa norit" (Aul. Gell., xiv. 6, 3). Cnf., too, Pliny, "Nat. Hift.," xxi. 4.

[2] W. E. Gladftone, "Juventus Mundi," p. 315; and fee Max Müller's "Effay on Comparative Mythology." ("Oxford Effays," 1856, p. 81.)

"With rofes crowned, on flowers fupinely laid,
Anacreon blithe the fprightly lyre effayed."

Love fleeping among the rofes and ftung by a
bee, or caught by the Mufes and bound with
wreaths of rofes, or the ode on "The Rofe,"
imitated by Dr. Broome, which begins:

"Come, lyrift, tune thy harp and play
Refponfive to my vocal lay ;
Gently touch it while I fing
The rofe, the glory of the Spring."

Thefe are famples of the feftive ideas connected
with rofes among the luxurious Afiatic Greeks.
The flavour of rofes was ufed to improve cookery,
and fo there was a Greek conferve, like our
marmalade, compofed of rofes and quinces.

In the Middle Ages, the rofe was one of the
few flowers which men found leifure to cultivate
in England. It would not be often feen on the
cottage-wall, as with us at prefent, but more frequently in the pleafance or even the little garden
on one fide of the caftle, fhut in between two of
its angles, fuch as may yet be feen at Stirling.
The French writer of the "Romaunt of the Rofe"
would naturally expect it to bloffom in the garden
which he fomewhat profanely, though only after
the fafhion of his time, defcribes as:

"There is no place in Paradife
So gode in for to dwell or be,
As in that gardin thoughtin me."

And the God of Love is attired by him in a
garment:

> "I purtraied and iwrought with floures
> By divers medeling of coloures ;
> Flouris there were of many gife,
> Ifet by compace in a fife.
> There lackid no of lure to my dome,
> Ne not fo much as floure of brome,
> Ne violet, ne eke pewinke,
> Ne flowre none that men can on thinke ;
> And many a rofe-lefe full long
> Was entermedlid there emong ;
> And alfo on his hedde was fet
> Of rofes redde a chapilet."[1]

A rofary is alfo defcribed—

> "Chargid full of rofis
> That with an hedge aboute enclofed is."

There "gretift hepe of rofes be;" and thefe "rofes redde" with their "knoppis," or birds, are dwelt on by the poet with the pleafure of a true rofe-lover.

But it is in Dante that the moft glorious and devotional ufe of the rofe is found ; a ufe from which comes our expreffion a "rofe-window," to indicate a large circular cathedral window filled with ftained glafs reprefenting faints and martyrs radiating from the central effulgence of Divine glory. Thus in the "Paradifo," he writes:

> "Lume è laffù, che vifibile face
>   Lo Creatore a quella creatura,
>   Che folo in lui vedere ha la fua pace ;
>
> "E fi diftende in circular figura
>   In tanto, che la fua circonferenza
>   Sarebbe al Sol troppo larga cintura.
>
>   *   *   *   *
>
> "Nel giallo della rofa fempiterna
>   Che fi dilata, rigrada e redole
>   Odor di lode al Sol che fempre verna."

---

[1] Anderfon's Poets, vol. i., pp. 281, 287.

And again, in the next canto:

> " In forma dunque di candida rofa
>   Mi fi moftrava la milizia fanta,
>   Che nel fuo fangue Crifto fece fpofa ;"

while the angel hoft, like bees humming round a rofe,

> " Nel gran fior difcendeva, che f'adorna
>   Di tanti foglie, e quindi rifaliva
>   Là dove il fuo amor fempre foggiorna.

> " Le facce tutte avean di fiamma viva,
>   E l' ale d' oro, e l' altro tanto bianco
>   Che nulla neve a quel termine arriva.

> "'Quando fcendean nel fior, di banco in banco
>   Porgevan della pace e dell' ardore,
>   Ch' egli acquiftavan ventilando il fianco.

> *   *   *   *

> " Chè la luce divina è penetrante
>   Per l' univerfo, fecondo ch' è degno,
>   Si che nulla le puote effere oftante."[1]

Surely no uninfpired writer ever penned fuch words of fplendid adoration and infight! The vifion may fitly clofe with the ftrain of another great thinker:

> " All is beauty,
> And knowing this is love, and love is duty ;
> What further may be fought for or declared ?"[2]

---

[1] " Paradifo," Canto xxx. 100 ; xxxi. 1-24.
[2] Browning, " The Guardian Angel."

CHAPTER XIV.

WOLVES.

THE wolf, as being univerfally dif-tributed, is fo well known that a large body of curious learning has grown up with it. Its tail is ftraight; which feems to eftablifh a ftructural difference between it and the numerous varieties of the dog. Yet naturalifts, fuch as the late Mr. Bell, have derived all dogs from the wolf, although Linnæus defcribes the former animal as "caudâ finiftrorfum recurvatâ." The Old World wolves are probably not fpecifically different from thofe of the New. They are found all over the Continent, and range from Egypt to Lapland. The jackal, a near congener, appears only in Eaftern Europe, while a variety known as the black wolf (*C. Lycaon*) is found in the Vofges Mountains, in the Alps, and the Pyrenees. As for the derivation of the word "wolf," its "fuggefted connection with Lat. 'vulpes,' a fox, is not generally accepted."[1] The

[1] Skeat, "Dictionary."

Sanfcrit form of the word is "vrika," the "tearer," or "render." In Icelandic it is "úlfr," whence our "wolf;" as "old" has become "wold." With the Northmen, the wolf was facred to Odin, who was always accompanied by two of thefe animals, Geri and Freki, which were fed with his own hand. At leaft two place-names in Lincolnfhire, Ulceby and Uffelby, retain traces of the wolf's Norfe name ;[1] while Wolverton, Woolmer, and the like, fhew that the Saxons alfo left their name for the creature in the local nomenclature of the country.

The wolf was in later hiftorical times the largeft wild beaft known to the Greeks; although, in the time of Xerxes, lions had fallen upon his baggage animals in Theffaly. It was regarded by them as the type of a bloodthirfty ravening creature, and as fuch frequently appears in Homer.[2] Its fkin was occafionally worn as a helmet, like the bearfkins of our troops. The Thracians, who joined the army of Xerxes, each bore two fpears, ufed for wolf-hunting, as arms. As being ftrictly a nocturnal animal, moft often feen in what was called "wolf-twilight," or grey dawn, the wolf was celebrated with the ancients in witchcraft and fuperftition. Homer places it together with the lion in the landfcape round the abode of Circe. Together with the Romans, it was an article of folk-lore among the Greeks that if a wolf faw a

---

[1] Streatfeild, "Lincolnfhire and the Danes," 1884, p. 72.
[2] Thus the Greeks and Trojans, mutually inflamed with rage, rufh upon each other "like wolves" ("Il.," iv. 471).

perfon firft, that man was ftruck dumb. So Plato makes Socrates fay, when angrily accofted by the fophift Thrafymachus: "I was difmayed and feared as I looked at him; and I verily believe, unlefs I had feen him firft, that I fhould have been ftruck dumb."[1] So "to fee a wolf," "wolf's wings" (like " pigeon's milk "), and "the wolf marrying the lamb," with others of the fame kind, became ufual Greek proverbs. Dean Trench juftly ftigmatizes " one muft howl with the wolves" as being the moft daftardly of all proverbs. This, however, is not due to Greek imagination.

The Egyptians fpecially affociated the wolf with the world of darknefs. It is reprefented on the painted walls of their catacombs and temples, and was probably connected by the priefts with fome efoteric doctrine of the tranfmigration of fouls. Wolf mummies are found at Ofioot, the ancient Lycopolis.

At Rome, the wolf, fuitably to the national character, was held in high honour. This took its rife from the fhe-wolf which had fuckled Romulus and Remus. Lupa, as Livy terms her, was the wife of Fauftulus, the royal herdfman; but fhe was

[1] "De Rep.," 336 d. Cnf. Virgil,

"Mærin lupi videre priores."

("Ecl.," ix. 54, and Theoc. xiv. 22.) "The ground or occafional original hereof was probably the amazement and fudden filence the unexpected appearance of wolves do often put upon travellers. But thus could not the mouths of worthy martyrs be filenced, who being expofed not only unto the eyes, but the mercilefs teeth of wolves, gave loud expreffions of their faith, and their holy clamours were heard as high as heaven."—(Sir T. Browne, " Vulgar Errors," iii. 8.)

foon deified under the title of Luperca, while the Lycean Pan's feftival (fo called becaufe he kept off wolves) was entitled Lupercalia, and was one of the moft popular of the old Roman feftivities. From the ftory connected with the birth of the founder of the city, the wolf was deemed facred to Mars. A clufter of Roman proverbs attached itfelf to this animal. " Lupus in fermone " was applied to any fudden appearance of the perfon who was being fpoken of at the time. " To have a wolf by the ears," meant to be in a fituation of great difficulty, from which advance or retreat was dangerous. " To fnatch the lamb from the wolf," " to fet the wolf over the flock," and the like, are famples of thefe proverbs. The reprefentation of the wolf, fometimes with, fometimes without the twin children, was a favourite device on Roman coins. It appears alfo on one of Ilerda. Art and poetry drew Romulus as rejoicing

> " Lupæ fulvo nutricis tegmine."

Among the magnificent imagery of the fhield worked by Vulcan and given by Venus to Æneas, we may be fure that thefe infant glories of the State were not forgotten :

> " Fecerat et viridi fœtam Mavortis in antro
> Procubuiffe lupam ; geminos huic ubera circum,
> Ludere pendentes pueros, et lambere matrem
> Impavidos ; illam tereti cervice reflexam
> Mulcere alternos et corpora fingere lingua."[1]

Dryden has caught much of the beauty of thefe lines :

[1] " Æn.," viii. 630.

" Here in a verdant cave's embowering ſhade,
  The foſtering wolf and martial twins were laid ;
  Th' indulgent mother, half reclined along,
  Looked fondly back, and formed them with her tongue,
  While at her breaſt the ſportive infants hung."

Ornytus is alſo pictured by Virgil as wearing a wolf-ſkin head-dreſs :

" Caput ingens oris hiatus
  Et malæ texere lupi cum dentibus albis."[1]

Ariſtotle evidently knew a good deal about the habits of the wolf. It produces blind puppies like a dog, he ſays. A pleaſant fable has attached itſelf to wolves, that they all produce young in a certain twelve days of the year, becauſe in ſo many days they once conducted Latona from the Hyperboreans to Delos, ſhe having changed herſelf into the form of a ſhe-wolf from fear of Juno. This ſtatement, however, he adds, ſeems to be as mythical as the ſtory that they only bear young once in their lives. They always live on fleſh, except when ailing, and then, like dogs, they eat graſs. Thoſe which lead a ſolitary life are more ready to eat men than thoſe which hunt in packs. In exceſſive hunger they will ſtoop to eat earth. Clearly Ariſtotle had ſifted much of the popular knowledge, as was his wont; but it is not ſurpriſing that he ſtates more of wolves than experience warranted.[2]

Pliny, on the contrary, although he lived ſo much later, was an eager liſtener to all old women's tales. The fat of wolves was eſteemed,

[1] " Æn.," xi. 680.
[2] " De Nat. Animal.," vi. 29 ; viii. 7.

he writes, above all. " New-wedded wives were
wont upon their marriage-day to anoint the fide-
pofts of their hufbands therwith at their firft
entrance, to the end that no charms, witchcrafts,
and forceries might haue power to enter in."
Again: " The muffle or fnout of a wolfe, kept
long dried, is a counter-charm againft all witchcraft and forcery; which is the reafon that they
ufually fet it upon gates of countrey ferms. The
fame force the very fkin is thought to haue which
is flaied whole of itfelf, without any flefh, from
the nape of the neck. And, in truth, ouer and
aboue the properties which I haue reported already
of this beaft, of fuch power and vertue it is, that
if horfes chance to tread in the tracts of a wolfe,
their feet will be immediately benummed and
aftonied. Alfo their lard is a remedy for thofe
who are empoifoned by drinking quickfiluer."
Some parts of the animal he prefcribes to be
mixed with Attic honey, as this is " fingular for
thofe whofe fight is dim and troubled." Likewife certain bones are found in wolves " which, if
they be hanged about the arme, do cure the
collicke." But his credulity was not yet fated.
" To come unto leechcraft belonging unto beafts,
it is faid that wolves wil not come into any lord-
fhip or territory, if one of them be taken, and
when the legs are broken, be let bloud with a knife
by little and little, fo as the fame may be fhed
about the limits or bounds of the faid field, as he
is drawne along, and then the body be buried in
the very place where they began firft to dragge

him. Others take the plough-ſhare from the plough wherewith the firſt furrow was made that yeare in the field, and put it upon the fire burning vpon the common hearth of the houſe, and there let it lie untill it be quite conſumed; and look how long this is in doing, ſo long ſhal the wolfe do no harm to any liuing creature within that territorie or lordſhip."[1]

Shakeſpeare, who has remembered to add "the tooth of wolf" to the hell-broth of his witches' caldron, had good reaſon for the ſelection, as this animal enjoyed an unenviable reputation in witchcraft. By the wondrous herbs of Pontus, the lover in Virgil was enabled to ſee Mœris turn into a wolf, and hide in the woods and call forth ghoſts from their ſepulchres,[2] that is, become a werewolf. This is the firſt mention in Latin literature of the *verſipellis* or turnſkin, but it ran through the magical authors. In Greece the ſuperſtition was well known ; certain Scythians near the Black Sea paſſed for wizards, becauſe once a year they became wolves for a few days, and then returned to their true form. The old Northmen fancied that by wearing coats of wolf-ſkin, men could become wolves at pleaſure. Indeed, the ſuperſtition has ſpread widely, and is at preſent largely believed among the Northern nations. In Germany the change is now effected by unclaſping or cutting a girdle made of the ſkin of a man who has been hanged, and faſtened by a buckle having ſeven

---

[1] Pliny, "Nat. Hiſt." (Holland), xxviii. 9, 10 ; xiv. 20.
[2] "Ecl.," viii. 97.

tongues. Trials of alleged were-wolves (*loup-garous*) were as numerous in France, during the fixteenth century, as were trials for witchcraft in Scotland. There are many traces of the belief in Ruffian folk-lore, and the wolf in the ftory of "Little Red Riding Hood" was probably a were-wolf.[1]

Before the age of Jupiter, wild beafts and ferpents were innocuous, faid the Latin poet:

"Ille malum virus ferpentibus addidit atris,
Prædarique lupos juffit."

And in his picture of peftilence devaftating a country, with much fkill he introduces the wolf:

"Non lupus infidias explorat ovilia circum,
Nec gregibus nocturnus inambulat ; acrior illum
Cura domat."

A ftill more beautiful comparifon reprefents the wolf as endowed with confcience, and, mindful of his offences againft man, flinking off into the wilds.[2]

"Velut ille, prius quam tela inimica fequantur,
Continuo in montes fefe avius abdidit altos,
Occifo paftore, lupus, magnove juvenco,
Confcius audacis facti, caudamque remulcens
Subjecit pavitantem utero, filvafque petivit."

And the horror of the portents attending Cæfar's death is intenfified by the howling of wolves:

"Et altæ
Per noctem refonare lupis ululantibus urbes."[3]

---

[1] See a good chapter on this curious fuperftition in Kelly's "Curiofities of Indo-European Tradition and Folk-lore ;" cap. ix. (1863).

[2] "Georg.," iii. 537 ; i. 130 ; "Æn.," xi. 809.

[3] "Georg.," i. 486.

In faĉt, the wolf was an animal ſuited to Virgil's poetry, and kept in ſtore by him, ready for any imaginative emergency. So when Turnus has to be repreſented raging againſt the foe, he is compared to a wolf. Dryden by no means enters into the full beauty of the paſſage, which ſhould be read in the original:

> "So roams the mighty wolf about the fold,
> Wet with deſcending ſhowers and ſtiff with cold ;
> He howls for hunger and he grins for pain,
> His gnaſhing teeth are exerciſed in vain ;
> And, impotent of anger, finds no way
> In his diſtended paws to graſp the prey.
> The mothers liſten ; but the bleating lambs
> Securely ſwig the breaſt beneath the dams."[1]

After his ordinary faſhion, Ælian adds to the marvels of Pliny reſpeĉting the wolf. It cannot bend its head back, he aſſerts ; but muſt look ſtraight forwards. If it ſhould happen to tread on a flower of the ſquill, it is at once rendered torpid ; ſo foxes take care to ſtrew ſquills in the dens of wolves.[2] This animal has left its traces in our botanical names. The lycopodium is ſo called from its reſemblance to the dark circular cuſhion under the wolf's foot, while its upper ſurface was ſeen by the fanciful in the lycopus, or gipſy-wort. The gaping mouth of the wolf has left its popular impreſſion in the lycopis or bugloſs (wolf's-face).

Wolves go back to a great antiquity, for their bones have been found in the foſſil cave of Aurignac in France, in Kent's Hole, and elſewhere ; while

[1] " Æn.," ix. 59.　　　　[2] "De Nat. An.," x. 26.

they are faid to have been feen, fo lately as Elizabeth's reign, in Dartmoor and Dean Foreft.   An amufing writer, who travelled through Sutherlandfhire in 1650, fays: "Specially here never lack wolves more than are expedient."   For the hiftory of the wolf in England, the reader may be referred to Harting's "Extinct Britifh Animals," where much information on them is collected. He decides that the animal became extinct in England fometime in the reign of Henry VII. In Scotland, wolves lingered until the end of the feventeenth century, the laft being killed in 1743; while the laft was killed in Ireland in 1770, at all events after 1766.[1]   An old belief averred that wolves could not live in England.

If proverbial lore, witchcraft, and fuperftitions of many kinds claim the wolf as a ufeful animal, the fabulift would be put to fore ftraits were he deprived of its affiftance.   Æfop and his imitators generally draw the wolf as the imperfonation of tyrannical greed ; as in the fable of "The Wolf and the Lamb."   Occafionally it is ufed to teach mankind a moral leffon, as in that of the boy who called " Wolf! wolf!" when there was no wolf, and was finally torn in pieces for his deceit. Once, however, the better part of the wolf—its wild and free nature—is defervedly recognifed in the fable of "The Wolf and Dog," when the latter tries to cajole the ftarving wolf to give up its freedom :

> " Be complaifant, obliging, kind,
> And leave the wolf for once behind."

---

[1] "Extinct Brit. Animals," p. 204.

But on the wolf unluckily feeing the collar round his friend's neck, then,—

> "He ſtarts and without more ado,
> He bids the abjeᴄt wretch adieu.
> 'Enjoy your dainties, friend ; to me
> The nobleſt feaſt is liberty.
> The famiſhed wolf upon theſe deſert plains,
> Is happier than a fawning cur in chains.'"

Vanière, the Jeſuit, in his "Prædium Ruſticum" (lib. xvi.), deſcribes in poetic language the capture of wolves in pitfalls, and then names a curious method of capturing them, viz., by the uſe of fiſh-hooks:

> "Mira frande lupum capies, piſcaria celans.
> Æra cibis ; carnes et inextricabile ferrum
> Hauſit ubi, vis nulla poteſt exſolvere rubras,
> Non ovium jam cæde ſuo ſed ſanguine fauces."

After his faſhion, Goſſon (1579), in order to help the Lord Mayor of London " to ſette his hand to thruſt out abuſes," drags in a ſimilitude from wolves which he muſt have found in ſome old author, but which has eſcaped us : "The Thracians, when they muſt paſſe over frozen ſtreames, ſende out theyr Wolues, which laying theyr eares to the yſe [ice], liſten for noyſe. If they hear any thing, they gather that it mooues ; if it mooue, it is not congealed. If it be not congealed, it muſt be liquide. If it be liquide, then will it yeelde ; and if it yeelde, it is not good truſting it with the weight of their bodyes, leſt they ſincke. The world is ſo ſlippery that you are often inforced to paſs over yſe. Therefore I humbly beſeech you to try farther

and truft leffe: not your Wolues, but many of
your Citizens haue already fifted the daunger of
your paffage, and in fifting beene fwallowed to their
discredite."[1]

[1] Stephen Goffon's "Schoole of Abufe" (ed. Arber), p. 56.

CHAPTER XV.

ANCIENT FISH-LORE.

"Vera vulgi opinio, quidquid nafcatur in parte naturæ ulla, et in mare effe, præterque multa quæ nufquam alibi."—(PLINY, *Nat. Hift.*, ix. 1.)

IN no department of natural hiftory is the ignorance and credulity of ancient writers fo noticeable as in their account of fifh. Our own popular mifconceptions with regard to the habits and economy of fifh may well induce us to view with indulgence the fhort-comings of ancient naturalifts; and the Fifheries Exhibition of 1883 feems to have effected but little improvement in this refpect. The knowledge of the people with regard to fifh, however, has increafed wonderfully between the reign of Henry VII. and our own days; in the cafe of ancient fcientific writers—ufing the word "fcientific" of the beft knowledge of the time— not only does the knowledge of fifhes and their economy appear not to have improved at all in the four hundred years which intervened between

Ariftotle and the elder Pliny, but it has abfolutely retrograded. Pliny believes more fables, and recounts with grave face more marvels than did the elder natural hiftorian, while he is not nearly fo difcriminating, and does not exhibit the fame common-fenfe as did his forerunner. The vaft-nefs of his own compilations, and his perpetual induftry in noting any circumftances of intereft connected with natural hiftory, fmothered his judgment. He had neither time to fift facts nor to weigh the authority to be attached to ftate-ments of other authors; and thefe defects leave his great " Natural Hiftory" a *rudis indigeftaque moles*, which compares unfavourably with the more exact and painftaking work of Ariftotle. He, on the contrary, muft have ftudied fifh practically, fo far as actual ftudy of natural hiftory was poffible in the judgment of his time, and betrays no fmall acquaintance with the claffification of fifh, and the differences which mark them off from quadrupeds and birds. Thus he divides them into fifh which produce young by eggs, like ordinary fifh, or fifh which produce their young alive—fifh which we now know to refemble quadrupeds in poffeffing warm blood, fuch as whales, dolphins, τὰ σελάχη, and the like. On their generation he was very well informed. Pliny, on the contrary, in addition to the ftatements of previous writers and of his own coadjutors, might have never feen a fifh fave fuch as appeared at his table. The migrations of fifh, whereby the moft ufeful families are brought at certain feafons annually to our fhores—tunnies, mackerel, and the like—had been inveſtigated by the Greek philoſopher. He had alſo learnt that this united movement of certain kinds of fiſh (οἱ χυτοί, as he terms them ; " fiſh that ſwim in companies ") was preliminary to their ſpawning near the coaſts in ſhallow water,[1] although his reaſons for theſe migrations might furniſh a logician with inſtances of the fallacy, *Non cauſa pro cauſa.* " Now of fiſhes," he remarks, " ſome migrate to the land from the ſea, and to the ſea again from the land, in order to avoid the extremes of heat and cold. Thoſe which are taken near the ſhore are better than oceanic fiſhes, for they have more, and better, ſuſtenance ; as wherever the ſun ſtrikes it produces more numerous, and better, and more tender creatures, juſt as may be ſeen in garden produce."[2] Poſſeſſing a wide knowledge, too, of the different modes of generation among fiſh, even he is not ſuperior to many prejudices, and to the influence of much which would now be termed folk-lore. " Some fiſh are ſprung from mud and ſand, even among ſuch families as generate in the ordinary manner with eggs. This happens in marſhes and ſuch places, juſt as is ſaid once to have happened at Cnidus. There the water was dried up by the dog-days, and all the mud taken out ; but the water began to teem with life as ſoon as the firſt ſhowers fell, and in this place little fiſh were generated as the water began to riſe." This is ſtill a vulgar belief. Another,

---

[1] Ariſtot., " De Nat. Anim.," v. 9.
[2] *Ibid.*, viii. 15.

which refembles the popular ftories of fhowers of frogs or fifh, is alluded to in the following words on the fifh called *aphye*: "They are produced in fhady and marfhy places when, after a period of fine weather, the earth has taken in much warmth, as is the cafe about Salamis and Marathon. In fuch places, then, the *aphrus* is produced in funny weather. In fome places alfo it is born, whenever much rain has fallen from the fky, in the foam (*aphrus*) which floats on the furface of the rain-water; and fometimes," he goes on to ftate, "it fprings from the foam on the furface of the fea." Here, probably, for the fake of etymology, he identifies the *aphye* (ἀ-φύω) and the *aphrus* (foam). Endlefs fables are told about the generation of eels at the prefent day. They find their prototype in the firft natural hiftorian. This kind of fifh, too, he fays, is not born from eggs or the ordinary generation of fifhes; and it is clear that this is fo from the fact that, when marfhes have been drained and the mud fuffered to harden, eels have appeared with the firft fhower; "but in droughts and lakes always full of water they are not generated, for they both live and are fprung from the water of fhowers." Nor do they fpring from worms, as fome think, "but from what are called the vitals of the earth, which of their own accord acquire confiftency in the mud and damp ground." And they are generated wherever there may be putre-faction in the fea and rivers—in the fea where the feaweed is thick, and round the edges of lakes and rivers, for there the heat prevails moft to cauſe putrefaction.[1]   A moment's reflection ſhews how ſimilar are the beliefs of labourers, and even of many in higher ſtations at the preſent day.

A queſtion has often been raiſed whether fiſhes ſleep. Ariſtotle has no heſitation in anſwering it in the affirmative. They do not, indeed, cloſe their eyes; but their motionleſs ſtate, ſaving the ſlow movement of the tail, proves it. In this ſleep he knew that they could be taken out by the hand or ſtruck with a ſtick. The tunny-catchers, too, while the tunnies are aſleep, are enabled to throw their nets around them. " The dolphin and whale, and ſuch as have an air-paſſage, ſleep on the ſea with their air-paſſage projecting through which they breathe, gently moving their fins; ere now ſome have heard a dolphin ſnoring."[2]   To ſome ſuch fable Milton alludes in his grand lines:

> "that ſea-beaſt
> Leviathan, which God of all His works
> Created hugeſt that ſwim the ocean ſtream ;
> Him, haply, ſlumb'ring on the Norway foam,
> The pilot of ſome ſmall night-founder'd ſkiff
> Deeming ſome iſland, oft, as ſeamen tell,
> With fixed anchor in his ſcaly rind,
> Moors by his ſide under the lee, while night
> Inveſts the ſea, and wiſhed morn delays."[3]

Among the ſingular fiſh which Ariſtotle knows and deſcribes may be named the angler, or fiſher-frog (*lophius piſcatorius*), and the electric ray (*raia torpedo*). The habits of life of theſe are detailed, juſt as modern ſcience knows them: the firſt, with the tempting baits at the end of the long

[1] Ariſtot., "De An. Hiſt.," vi. 14, 15.
[2] *Ibid.*, iv. 10.   [3] "Par. Loſt," i. 200.

line-like proceffes on its head, while itfelf lies concealed in the fand; the fecond, with its powerful
natural battery, by which it ftuns fifh before it
feizes them.   He alfo mentions that it has the
power to benumb men, as our modern fifhermen
fometimes find to their coft.[1]   The anthias, when
taken, endeavours to faw the line off on the rocks,
juft as falmon do, when hooked in a Scotch ftream,
with ledges of flate.   The fcolopendra has an eafy
mode of efcaping the hook.   When it has
fwallowed one it turns infide out, and, fo having
rejected the hook, turns back again.   The fox-fifh
has another device : it choofes the line above the
hook for attack, bites it through, and fo efcapes ;
but night-lines fet with many hooks prove fatal
to this fifh.   Of the *glanis*, as he calls it—that
is the *filurus*—Ariftotle tells a ftory which has
actually been proved true in the cafe of the
common male Englifh ftickleback (*gafterofteus
trachurus*), which thus acts as guard to its neft,
and will not allow a female to approach the eggs.[2]
" Of river fifhes, the male glanis takes great care
of its young.   The female, having brought them
into exiftence, departs ; but the male, noting
where moft of the fpawn adheres, acts as guardian
of the eggs, and continues to do fo, warding off
the other little fifh left they fhould deftroy the
brood.   And this it does for forty or fifty days,
until the brood has grown and is able to efcape

<hr/>

[1] For the ftatements contained in this fection fee a curious
chapter (ix. 25).
[2] See Yarrell, " Hift. of Britifh Fifhes," ii., p. 77.

from other fiſh. This circumſtance is known by
the fiſhers from the faɛt of the fiſh moaning and
uttering a roar when it keeps off intruders."
With the exception of this latter marvel, the
procedure of the glanis is preciſely that of the
ſtickleback. Although Ariſtotle has miſtaken the
fiſh, the obſervation is acute, and ſhews how much
the philoſopher was in advance of his age. The
habits of the ſepia, in diſcharging its ink, were alſo
familiar to him. A paragraph reſpeɛting the
poulpe will ſhew the ſingular manner in which
faɛt and fable are mingled with the ſtatements of
, even the beſt of ancient naturaliſts: "Now the
polypus is a fooliſh creature, for it will come to a
man's hand if he puts it into the water; yet it is
a creature of ſome contrivance, for it colleɛts all
its prey into the den where it lives, and, when it
has conſumed the moſt uſeful parts, it caſts out the
ſhells and fragments of the crabs and ſea-ſnails and
the ſpines of the little fiſh, and chaſes the fiſh
which then come together to them, changing its
colour, and adapting itſelf in hue as much as
poſſible to the ſtones around. It adopts the ſame
device when terrified." He is ſomewhat narrow
in his views in a ſucceeding ſentence: "Among
fiſh, the *rhine*" (ſeemingly a kind of ſhark) "is the
only one to change colour like the polypus." This
is probably a common device with moſt fiſh, and
is well known to be the caſe with trout. In Mr.
St. John's "Natural Hiſtory and Wild Sports of
Moray," ſome ſingular inſtances are related of this

power in trout to affimilate their colour to their furroundings.

If Ariftotle contains many facts with not a few fables, Pliny's " Natural Hiftory of Fifh " confifts of many fables with but few facts. He is omnivorous and indifcriminating ; like his own Silurus, "a great devourer, and maketh foule work, for no living creatures come amiffe unto him ; he fetteth up all indifferently." Marvels of every kind are dear to him, fuch as the Indian fifhes, like eels, fixty cubits long, and fo ftrong that when elephants come to the river to drink, they catch their trunks with their teeth, and " mauger their hearts, force them downe under the water." A few more fpecimens of his curioufly blended facts and fancies may be given. All fifh fuffer much from cold, " but thofe efpecially who are thought to have a ftone in their head, as the pikes, the chromes, fcienæ and pagri." Again, " The Arcadians make wonderous great account of their exocœtus, fo called for that hee goeth abroad and taketh up his lodging on the dry land to fleep." Ariftotle was inclined to be credulous when treating of eels. Liften to Pliny : " Yeeles live 8 yeares. And if the North wind blow they abide alive without water 6 daies, but not fo long in a Southern wind. Of all fifh, they alone if they lie dead, flote not above the water." The whole life-hiftory of the eel is ftill fuch an enigma that readers muft be cautious how they fmile at Pliny's ftories. Take the following for inftance : " There is a Lake in Italy called Benacus,

through which the river Mincius runs; at the
iffue whereof everie yere about the moneth of
October, when the Autumne ftar Arcturus,
whereby the lake is troubled as it were with a
winter ftorme and tempeft, a man fhall fee rolling
amongft the waves a wonderfull number of thefe
Yeels wound and tangled one within another; in-
fomuch as in the leapweeles and weernets devifed
for the nonce to catch them in this river, there be
found fometime a thoufand of them wrapped
together in one ball." After the merriment which
fuch a ftory is liable to excite has abated, it is
worth while turning to a book juft publifhed by a
fifherman who has carefully ftudied the habits of
eels in the Broads of Norfolk. " A very curious
phenomenon," he fays, " is fometimes obfervable
in the upper waters of the Yare and Waveney:
the eels come down in large folid balls from one
to two feet in diameter, heads infide and tails out;
and thefe living balls roll down the river, and
plump into the nets with fuch force as to carry
them away, for which reafon the eel-fifhers at the
mills dread their coming. We cannot even guefs
at the caufe of this fingular eel-freak."[1]

The Echeneis, of courfe, is fabled by Pliny to
ftay fhips; " for that caufe alfo it hath but a bad
name in matters of love, for inchanting as it were
both men and women. Moreover, it hath this
vertue, being kept in falt, to draw up gold that is
fallen into a pit or well, being never fo deep, if it is

[1] "The Broads of Norfolk," p. 216. Blackwood, 1883.
By G. C. Davies.

let down and come to touch it." Victor Hugo has thrilled numberlefs readers with his account of the huge poulpe that attacked a man, and many ftories, fabulous and otherwife, have in recent years been feen in print about the fize and fierce- nefs of poulpes and calamaries. Pliny gives a marvellous account of the killing of fuch a monfter, " whofe head was as big as a good round hogfhead or barrel that would take and contain 15 amphores." His words implicitly contain all the fabulous as well as the true recitals concerning thefe monfters which have appeared of late years. Much of Pliny's " Hiftory " is a tranflation from Ariftotle, with many fables and fcraps of Italian folk-lore appended. We muft own to ignorance of the *aries* or ram-fifh, which muft poffefs what our forefathers would have termed " a fhrewd nature," for it is " a very ftrong theef at fea, and makes foule work where he comes; for one while he fquats clofe vnder the fhade of big fhips that ride at anker in the bay, where he lies in ambufh to wait when any man for his pleafure would fwim and bath himfelf, that fo he might furprife them: otherwhiles he puts out his nofe above the water to fpie any fmall fifher boats comming, and then he fwimmeth clofe to them, overturneth and finketh them." His teaching on the generation of fifhes is marked with vague credulity. His anthias, too, cuts the line afunder with the fharp, faw-like fins which it bears on its back, while the fargons fret it in two againft a fharp rock. His laft chapter on fifh is delightful, and has been the ſource of many of the fabulous tales of later ages. Some fiſh are friendly, he tells, others hateful to each other: " The Mullet and the ſea-Pike hate one another, and be ever at deadly war ; likewiſe the Congre and the Lamprey ; inſomuch as they gnaw off one another's taile. The Lobſter is ſo afraid of the Polype, or Pourcuttell, that if he ſpie him neere, he evermore dieth for very woe. The Lobſters are ready to ſcratch and teare the Congre ; the Congres, again, do as much for the Polype. On the other ſide there be examples of friendſhip among fiſhes beſides thoſe of whoſe ſociety and fellowſhip I have already written, and namely between the great whale Balæna and the little Muſculus. For whereas the whale aforeſaid hath no uſe of his eies (by reaſon of the heavy weight of his eie-browes that cover them), the other ſwimmeth before him, ſerveth him inſtead of eies and lights, to ſhow when he is neere the ſhelves and ſhallows, wherein he may be ſoon grounded, ſo big and huge he is." This ſtory has greatly taken the fancy of many old Engliſh writers, and it is evidently capable of being largely moralized. For example : " The ancients give for an Hierogliph of a wiſe Senate and able Counſell a little fiſh going before the great whale, diſcovering ſhallows and other dangers, and ſhewing the way by the motion of itſelf. This living, the whale is ſafe, but being dead, he knoweth not what to do."[1]

---

[1] " Sion's Plea againſt the Prelacy." See, too, S. Goſſon's " Schoole of Abuſe " (ed. Arber), p. 55. The above citations

The many curiofities of fifh-life are often dwelt upon by mediæval and later writers. They alfo fancied that analogues of all things living on earth were to be found in the fea. Thus Walton writes of the wonders which the Tradefcants collected into their mufeum. This yet exifts under the name of their friend Afhmole, at Oxford: "You may there fee the Hog-fifh, the Dog-fifh, the Dolphin, the Cony-fifh, the Parrot-fifh, the Shark, the Poifon-fifh," and others. And we will follow his example in " fweetening this difcourfe out of a contemplation in divine Du Bartas," after duly cautioning readers that this poet's works, tranflated into Englifh by Sylvefter, form 670 folio pages of the moft extreme dulnefs imaginable:[1]

> "God quickened in the fea and in the rivers
> So many fifhes of fo many features,
> That in the waters we may fee all creatures
> Even all that on the Earth are to be found,
> As if the world were in deep waters drowned.
> For Seas—as well as Skies—have Sun, Moon, Stars ;
> As well as Air—Swallows, Rooks, and Stares ;
> As well as Earth—Vines, Rofes, Nettles, Melons,
> Mufhrooms, Pinks, Gilliflowers and many millions
> Of other plants, more rare, more ftrange than thefe
> As very fifhes, living in the feas ;
> As alfo Rams, Calves, Horfes, Hares and Hogs,
> Wolves, Urchins, Lions, Elephants and Dogs,
> Yea, Men and Maids," etc., etc.

Walton proceeds to enumerate, from Ælian and

---

from Pliny belong to "Nat. Hift.," ix. 15, 16, 19, 21, 22, 25, 30, 44, 50, 51, 59, 62 (Holland's Tranflation).

[1] "Du Bartas, His divine Weekes and Workes, with a Compleate Collection of all the other moft delightfull Workes, tranflated and written by y't famous Philomufus Jofuah Sylvefter, Gent." London, 1641.

Oppian, in whom any number of fimilar marvels may be found, fome of the moft curious fimilarities between fea and land creatures, the hermit, Adonis, and the like. The latter fifh finds much favour in his eyes, "becaufe it is a loving and innocent fifh, a fifh that hurts nothing that hath life, and is at peace with all the numerous inhabitants of that vaft watery element ; and truly, I think, *moft Anglers are fo difpofed to moft of mankind.*"[1]

Spenfer, who fwept everything into his verfe, was not unmindful of the refources of pifcine monfters offered him by the fea. They may amufe fifhermen, when, as his own Colin fays:

> "Sad winter welked hath the day,
> And Phœbus, wearie of his yearlie tafke,
> Yftabled hath his fteedes in lowly lay,
> And taken up his ynne in Fifhes hafke."[2]

And for his unknown pifcine terrors, they are not even furpaffed by the monfters of the deep which Schiller makes his Diver fee in the perilous plunge for the goblet. In truth, it is a gruefome catalogue:

> "Eftfoones they faw an hideous hoaft arrayd
> Of huge fea-monfters, fuch as living fence difmayd.

> "Moft ugly fhapes and horrible afpécts,
> Such as dame Nature felfe mote feare to fee,
> Or fhame, that ever fhould fo fowle defects
> From her moft cunning hand efcaped bee ;
> All dreadful portraicts of deformitee :
> Spring-headed hydres ; and fea-fhouldring whales,
> Great whirlpooles, which all fifhes make to flee ;
> Bright fcolopendraes armd with filver fcales ;
> Mighty monoceros with immeafured tayles ;

---

[1] See "Compleat Angler," part i.
[2] "The Shepheard's Calender," November.

> " The dreadful fifh, that hath deferv'd the name
> Of Death, and like him lookes in dreadful hew ;
> The griefly wafferman, that makes his game
> The flying fhips with fwiftnefs to purfew ;
> The horrible fea-fatyre, that doth fhew
> His fearefull face in time of greateft ftorme ;
> Huge ziffius, whom mariners efchew
> No leffe than rockes, as trauellers informe ;
> And greedy rofmarines with vifages deforme :

> " All thefe and thoufand thoufands many more,
> And more deformed monfters thoufand fold
> With dreadfull noife and hollow rombling rore
> Came rufhing, in the fomy waues enrold."

Soon afterwards Spenfer's travellers fee the five Sirens, as if he was determined that the fea fhould hold wonders enough. Thefe were once " faire Ladies," but now

> " Depriv'd
> Of their proud beautie, and th' one moyity
> Transform'd to fifh for their bold furquetry ;
> But th' upper halfe their hew retained ftill,
> And their fweet fkill in wonted melody."

The laft line, however, is worthy for its fweetnefs to compare with anything which even Milton wrote on mufic.[1]

From fabulous to the fifh of everyday-life is an eafy ftep. Another poet of the Elizabethan period fhall fum up the ftore of fifh with which Nature, niggardly in beftowing other charms, has enriched Lincolnfhire. The German Ocean was even in his time recognifed as the Mother of Wealth :

> " What fifh can any fhore or Britifh fea-town fhew
> That's eatable to us, that it doth not beftow

---

[1] Spenfer, " Faerie Queene," bk. ii. xii. 23, 31.

> Abundantly thereon ?   The herring, king of fea,
> The fafter-feeding cod, the mackerel brought by May,
> The dainty fole and plaice, the dab, as of their blood ;
> The conger finely foufed, hot fummer's cooleft food ;
> The whiting known to all, a general wholefome difh,
> The gurnet, rochet, mayd and mullet, dainty fifh ;
> The haddock, turbet, berb, fifh nourifhing and ftrong ;
> The thornback and the fcate, provocative among ;
> The weaver, which although his prickles venom be,
> By fifhers cut away, which buyers feldom fee,
> Yet for the fifh he bears 'tis not accounted bad ;
> The fea-flounder is here as common as the fhad,
> The fturgeon, cut to keggs (too big to handle whole),
> Gives many a dainty bit out of his lufty jole."

And much more to the fame import, often profaic
enough, and a warning to poets who commit themfelves to enumerations of natural objects.   We
will conclude with one more curious fuperftition
about the ofprey.   Drayton's lines prove that the
bird was fufficiently common in Lincolnfhire in his
time; though, alas ! it has now been long extinct,
.and the few that do crofs the county on migration
meet with the ufual fate of all rare birds, being at
once fhot and "fet up" in glafs cafes, lafting
emblems of the felfifh and wanton cruelty of their
captors :

> "The ofpray oft here feen, though feldom here it breeds,
> Which over them the fifh no fooner do efpie,
> But (betwixt him and them by an antipathy)
> Turning their bellies up, as though their death they faw,
> They at his pleafure lie to ftuff his glutt'nous maw."[1]

---

[1] Drayton's " Polyolbion," Song 25.

CHAPTER XVI.

MYTHICAL ANIMALS.

" Libri Græci miraculorum fabularumque pleni ; res inauditæ, incredulæ ; scriptores veteres non parvæ auctoritatis."—(AUL. GELLIUS.)

IN Greek and Roman literature, particularly in the earlier authors, many mythical beings are found, just as in the primitive history of almost all nations. Sometimes the philosophical reason for a belief in these mythical creatures is evident after a little consideration. Thus the numerous worms or serpents—many of which have left their trail on local names, and many more in the traditional folk-lore of England—are undoubtedly due to the old Norse reverence for these creatures; perhaps because, in the Scandinavian cosmogony, the earth was girdled by a monstrous serpent called Jörmungandr. Again, the numerous and fantastically-sized sacred fish of the Buddhists are referable to these devotees' fondness for fish; while the mythically-shaped creatures, peacocks, elephants, and the like, common in Oriental art, are but exaggerations of forms familiar to Eaſtern tribes from their infancy. In claſſical literature, the genius of the two nations delighted to exerciſe itſelf in the production of groteſque monſters, which fancy frequently inveſted with ſtriking attributes; and the poets, embalming theſe conceptions in their verſe, handed them on to numerous generations of writers and ſtudents of ancient Greece and Rome. Wordſworth has well pointed out that the natural features of Greece, when paſſed through the alembic of poetic fancy, at once reſulted in many a beautiful, many a monſtrous brood of ſupernatural creations :

> "The Zephyrs fanning, as they paſſed, their wings,
> Lacked not for love fair objects whom they wooed
> With gentle whiſper. Withered boughs groteſque,
> Stripped of their leaves and twigs by hoary age,
> From depth of ſhaggy covert peeping forth
> In the low vale, or on ſteep mountain-ſide ;
> And, ſometimes, intermixed with ſtirring horns
> Of the live deer, or goat's depending beard,—
> Theſe were the lurking Satyrs, a wild brood
> Of gameſome deities ; or Pan himſelf,
> The ſimple ſhepherd's awe-inſpiring god."[1]

Beſides the richneſs of native fancy, a large infuſion of Oriental beliefs coloured Greek mythology. It is exceedingly difficult to eſtimate the amount and value of theſe importations. Save in the " Odyſſey," Homer is comparatively free from them. There he ſeems intentionally to have dowered his verſe with much of the richneſs and many of the fantaſtic characteriſtics of the Eaſt.

[1] See " The Excurſion," pp. 134-139.

Phœnician failors and merchants brought into
Greece a ftock of marvels which they may have
gathered from fuch ftory-tellers as may yet be
heard in Bagdad, and read of in the pages of
the "Arabian Nights." Many of the fhipwrecks
of Odyffeus, the marvels of Circe's ifland, the
prodigies vifible to the hero in the Necyia, are of
a diftinctly Eaftern dye. The Orontes did not
flow alone into the Tiber; and tales of travellers,
always acceptable to ftay-at-home folk, came with
a natural fitnefs from the fertile lands of the Eaft
to the Weftern World. How greatly the Greeks
were indebted to the Egyptians for much of their
fyftem of divinities, and efpecially for fo many of
their conceptions of the future ftate, may be feen
in Herodotus. The fables of Charon and his
obole, of Cerberus, of the ftern Rhadamanthus,
and the like, are fpecimens of this mythology of
Hades. The worfhip of Aphrodite and Hercules
came to Greece from the Phœnician cult of
Aftarte and Melkarth. The revels connected
with the worfhip of Dionyfus were due to Egypt.

Over and above the fyftems of the greater
divinities which were elaborated by the Greeks
and Romans, they were exceedingly hofpitable to
the gods of conquered lands. Thefe were introduced with much of the ftrange ritual connected
with them, and large numbers of the vulgar were
carried away with their worfhip. Many ftrange
and grotefque conceptions of what may be termed
popular mythology alfo fucceeded in entering the
claffical lands—fome from one caufe, fome from

another. Thus Herodotus appears to have taken,
so Heeren fuppofes, a caravan journey through
North Africa, as defcribed by him in iv. 181-185;
and we can trace the marvels which were told him
in his journey becoming, on his return, part and
parcel of Greek thought. To this were due the
marvellous animals which his defcription of a large
ftrip of territory, being θηριώδης, weftward of the
river Triton, allowed the play of fancy at once to
create: oxen which fed backwards, owing to the
projection of their horns in front; fnakes, lions,
elephants, bears, afps, horned wild affes, dog-
headed apes, monfters with no heads and eyes in
their chefts, "as the Libyans tell, and wild men
and wild women, and multitudes of other creatures
in nowife fabulous," as the hiftorian feelingly fays.[1]
It is curious that the monftrous creatures which
Robinfon Crufoe met are placed by Defoe in this
region. Moft probably many of thefe reports
were induftrioufly fpread abroad by the Carthaginians to prevent troublefome neighbours from
interfering with their commerce; but much muft
be affigned to the tendency of all ignorance to
exaggerate. Here, too, was the country of the
Garamantes, Lotophagi, and others, where Greek
fancy could plant marvels of any kind; much as
our popular writers take New Guinea and the
Cannibal Iflands for the home of their ideal
monfters.

Modern philology has done much to winnow
the corn from the chaff in thefe mythological

[1] Herod., iv. 191.

speculations. It is now generally recognised that astronomical phenomena, the succession of day and night, the procession of the sun through the signs of the zodiac, and the like, underlie many of the most grotesque of these classical beliefs. " By a succession of the most fortunate circumstances, the astronomical books of three of the principal religions of the ancient world have lately been recovered—the Veda, the Zend-Avesta, and the Tripitaka. But not only have we thus gained accefs to the most authentic documents from which to study the ancient religion of the Brahmans, the Zoroastrians, and the Buddhists, but by discovering the real origin of Greek, Roman, and likewise of Teutonic, Celtic, and Slavonic mythology, it has become possible to separate the truly religious elements in the sacred traditions of these nations from the mythological crust by which they are surrounded, and thus to gain a clearer insight into the real faith of the ancient Aryan world."[1] It may, however, be reasonably doubted whether the universal solvent of a solar myth has not been too frequently applied. Many of the mythological animals of the ancients appear to have been created for a moral purpose; therefore it is out of place to regard them as emblems of astronomical phenomena. " Upon deliberate consideration," says Lord Bacon, " my judgment is that a concealed instruction and allegory was originally intended in many of the ancient fables." And the least

[1] Max Müller, " Selected Essays " (Longmans, 1881), vol. i., p. 5.

reflection will ſhew to a believer in revelation that the Greeks often ſpake of things higher than they knew when they diſcourſed of mythical animals and events. Theſe ſtories are many of them waifs and ſtrays which have floated down the ſtream of time from the original home of the human race. They are part of the fairy-tales told in the nurſery of man during the infancy of the world, drawn by the Greeks and Romans from " the common ſtock of ancient tradition, and varied but in point of embelliſhment, which is their own. And this principally raiſes my eſteem of theſe fables ; which I receive, not as the product of the age or inven- tion of the poets, but as ſacred relics, gentle whiſpers, and the breath of better times, that from the traditions of more ancient nations came at length into the flutes and trumpets of the Greeks." [1] In purſuance of this view, Lord Bacon explains Typhon to mean a rebel ; Proteus, matter ; the Sphinx, ſcience ; the Sirens, pleaſures ; and Scylla and Charybdis, the middle way ; and ſo forth. The Cyclopes again, ſo poetically deſcribed by Virgil :

> " Centum alii curva hæc habitant at littora vulgo
> Infandi Cyclopes, et altis montibus errant ;"

and again :

> " Cernimus adſtantes necquidquam lumine torvo
> Ætnæos fratres, ... concilium horrendum,"

become, in his view, miniſters of terror aſſiſting a deſpotiſm. The poets, however, do not ſeem to

---

[1] Bacon's " Wiſdom of the Ancients," Preface.

bear him out in this interpretation; with them the Cyclopes rather reprefent the exceffive toil required in forging iron, and fhew that the bleffings of civilization are only attained by conftant and unenviable labours—" as when the Cyclopes haftily forge thunderbolts out of tough maffes of metal; fome take in and blow out the gales of heaven from their bellows of bullhide, others dip the hiffing bronze into the lake. Ætna groans at the weight of the anvils placed upon her. They, vying with one another with mighty force, raife their arms together, and turn with ftout-holding forceps the weighty iron."[1]

Kingfley opined that our own Teutonic fore-fathers imported their elves, trolls, pixies, and the like, from the heart of Afia. They feem to us ra⁺her a fpontaneous growth of the northern mind, fuited to the attributes of the "blamelefs Hyperboreans," who gave them birth. No monftrous brood are they, fwelling with envy and rage againft heaven and earth, like Hylæus, Typhoeus, and the remnants of the giants of Grecian fancy, but kindly houfehold fprites, will-ing to be friendly with man; and, if a little trickfy at times, eafily appeafed by a bowl of milk, a frefhly-baked cake, or the like. Even Thor and Odin (Thunder and Wind) were magnanimous and placable, if huge and all-powerful. Images of terror and fuperhuman force and cruelty naturally affected the Greeks in their beautiful land and mild, foporific climate. The Scandinavians, on

---

[1] Virgil, " Æn.," iii. 634, 677 ; " Georg.," iv. 170-175.

the other hand, with their barren cliffs, vaft precipices, and ftern lengthy winters, were more acceffible in the way of contraft to gentler and fofter beings who would refine the ruggednefs of their national character. Yet early northern art, like Greek poetry, played with and expanded its types of the fupernatural into a thoufand quaint interlaced devices. That Chriftianity underlaid moft of thefe curious carvings, fo familiar to admirers of Pictifh or Scandinavian ftone-fculpture, is manifeft from the circumftance of the crofs in fome floriated and interwoven pattern frequently forming the foundation of a wealth of ornamentation and imagery. In this crofs are often found fine boffes or holes (for the five wounds of our Lord), juft as is fo frequently obferved in the fine croffes of Cornwall. The imagination of the carvers was allowed to run riot round this fymbol of falvation. Among the moft lovely twifted cable-patterns are feen on the old Scotch ftones birds, fifh kiffing each other (as at Mortlach), deer purfuing each other (Elgin), horfe-headed fifh (Upper Manbean), ferpents, bulls, horfes, bears, fifh with the adipofe fin reprefented—proving how carefully the artift had copied nature—galleys, reindeer (near Grantown), wild boars with very confpicuous tufks, ofpreys eating fifh, and the like. " The eye," fays Burton, " becomes almoft tired with the endlefs fucceffion of grim and ghaftly human figures, of diftorted limbs, of preternatural beafts, birds, and fifhes, of dragons, centaurs, and entwined fnakes." The germs of this characteristically circular ornamentation may be seen in the singular curves and circles of early Celtic and prehistoric times, many of which are still preserved in stone. And yet there are points to connect Scandinavian with Oriental stone-sculpture. Even in Ceylon stones may be seen with elephants, crescents, serpents, and geese carved on them. At Canna, in the Hebrides, in a little churchyard, a broken cross of yellow sandstone exists bearing curious carvings, and, among other things, exhibiting a camel,[1] "the only instance of it known in Scotland." Lions are also, at least twice, found among the creatures carved on the sculptured stones of Scotland. Such marvellous kinship is there between the different families of the human race; so curiously have early beliefs expanded, shrunk, disappeared, and again emerged in the most unexpected localities.

Another fertile source of zoological myths among the ancients was their total ignorance in many cases, in others what is equally dangerous, their little knowledge of natural history. Pliny knew a little about the cuckoo, for instance; but, trading on this, he simply invented the fable that it is eaten by its own kind. This tendency may often be seen in his recitals. The sea-monster, κῆτος, was idealized from the large sharks of the Mediterranean by the help, in all probability, of

[1] "The Hebrides," by Miss Gordon Cumming, 1883, p. 112. See also "Rude Stone Monuments," by Fergusson, *passim*; "The Sculptured Stones of Scotland" (Spalding Club), 2 vols. fol., by John Stuart, 1856-67; "The Bookhunter," J. H. Burton, p. 396 (Edinburgh, 1863).

Phœnician traditions of whales; and then the next ftep was eafy, the ftories of Andromeda and Hefione, and their releafe by heroes. Similarly the hydra[1] was magnified from the fnake. The Harpies, too, which inhabited the Strophades, were faint fhadows of travellers' tales from the Eaft. Large bats were fpeedily transformed by credulous wonder into

> " Virginei volucrûm vultus, fœdiffima ventris
> Proluvies, uncæque manus, et pallida femper
> Ora fame ;"

and then the poet may well add, " Triftius haud illis monftrum."[2]

The procefs of mythological creation can be feen in the "Odyffey," where the word Harpy firft occurs. In it Harpies are fimply ftorm-winds which fweep off their victims; the fouler features were afterwards added.[3] Once more, the griffin was fabled to poffefs a lion's body, with an eagle's face and wings. When we are told that it was faid to guard the gold-mines in the country of the Arimafpi, we are at no lofs to difcover the reafon which prompted its creation.[4] It is not fo eafy to trace the genefis of the Chimæra, " the invincible Chimæra," as Homer terms it, " which was of divine, and not of mortal lineage, a lion in front, a dragon behind, and a fhe-goat in the midft, breathing forth the dreadful might of blazing

---

[1] " Quinquaginta atris immanis hiatibus Hydra
  Sævior intus habet fedem."—Virgil, " Æn.," vi. 576.
[2] Virgil, " Æn.," iii. 214, and 223 *feq.*
[3] " Odyffey," xx. 77.
[4] See Virgil, " Ecl.," viii. 27, and the note of Forbiger.

fire."[1]   At all events, it ſerved Virgil for an objeƈt
on which to expend his imagination when it
figured on the helmet of Turnus:

> " Cui triplici crinita juba galea alta Chimæram
> Suſtinet, Ætnæos efflantem faucibus ignes ;
> Tam magis illa fremens, et triſtibus effera flammis,
> Quam magis effuſo crudeſcunt ſanguine pugnæ."

And the Laureate was probably indebted to it for
the fine imagery of his hero Arthur's helmet which
Guinevere ſaw,

> " Wet with the miſts and ſmitten by the lights,
> The Dragon of the great Pendragonſhip
> Blaze, making all the night a ſtream of fire."

Occaſionally the poets, and eſpecially the ſyſ-
tematizers of the national theology, from one
monſter fabled the birth of others.   Thus from
Typhoeus and Echidna, Geryon, Orthos, Cerberus
and the Hydra were ſaid to have ſprung.
Naturally, this principle was capable of indefinite
expanſion in the hands of imaginative writers.
Natural but unfamiliar objeƈts ſupplied the
nucleus round which other myths might centre.
Thus the aſtoniſhment of their neighbours when
they firſt beheld the Theſſalians mounted on
horſeback led to the formation of thoſe fabulous
creatures, the Centaurs.   The Greeks, it is well
known, at the ſiege of Troy were unacquainted
with the art of riding.   Again, the ſight and
ſound of a roaring whirlpool, with much broken
water and ſurf, furniſhed the hint for ſome ſea-
ſong, which told of Scylla and her ſix heads, each

[1] " Il.," vi. 179 ; and " Æn.," vii. 785.

poſſeſſed of three rows of teeth, while below the waiſt ſhe developed into frightful dogs, which never ceaſed barking. Then the poets amplified to their own liking. Thus Homer, "the great father of them all:"

"Now, in the middle of the cliff is a darkling cavern, looking weſtward, turned towards Erebus, nor in ſooth could a vigorous man from a hollow ſhip having ſhot an arrow penetrate with it the depths of that hollow cave. Therein dwells Scylla, barking terribly. Her voice is like that of a young whelp, and herſelf is, in truth, a monſtrous woe; nor would anyone rejoice when he beheld her, nor even a god, if he approached her. All her twelve feet are miſ-ſhapen, and her ſix necks are very long, and ſet on each is a terrible head, with three ranks of teeth in it, many and crowded together, full of black death. Up to the midſt of her ſhe is ſunk in the hollow cavern, but thruſts out her heads from its dreadful gulf. And there ſhe fiſhes, gazing round the cliff for dolphins, and ſea-dogs, and any greater monſter which ſhe can ſeize, whereof deep-voiced Amphitrite tends many thouſands."[1] The undefined horror of much of this deſcription largely enhances its terror. What, for inſtance, is more ſtriking than the expreſſion, "teeth full of black death"? Had Virgil been contented with his Scylla, and the cliffs reſounding with blue ſea-dogs, his monſter would have gained in vaſtneſs and awe; but he muſt needs particularize, and at once the charm of the "monſtrum,

[1] Hom., "Od.," xii. 89.

horrendum, informe, ingens," difappears.   " A
cave reftrains Scylla with its dark receffes as fhe
thrufts forth her mouths and drags fhips on to the
rocks.   Above, fhe bears the countenance of a
man, and as far as her loins is a virgin, with
beautiful breafts; her extremities form a fea-
monfter" (*piftrix*), "with huge body and the
womb of wolves attached to the tails of dolphins."[1]
A pretty account of the trahsformation of Scylla
into this fea-monfter may be found in Ovid
(" Met.," xiv. 60-67).   The poetic inftinct, however, is ftrong with Virgil ; when defcribing the
defcent of two Centaurs from fnowy mountains he
refrains from particulars, and merely calls them
" nubigenæ "[2]—cloud-fprung.   Though primarily
denoting their parentage, the epithet is in other
ways a happy one from its indefinitenefs.

Even fifty years before the Chriftian era, not
only the monftrous creatures above fpoken of, but
alfo the ordinary deities, were only believed in by
the vulgar.   Philofophers, however, either tacitly
endured or treated them with open contempt.
" The very children and old women ridiculed
Cerberus and the Furies, or treated them as mere
metaphors of confcience.   In the deifm of Cicero,
the popular divinities were difcarded, the oracles
refuted and ridiculed, the whole fyftem of divina-
tion pronounced a political impofture, and the
genefis of the miraculous traced to the exuberance
of the imagination, and to certain difeafes of the
judgment."[3]   Comedy at Athens early learnt to

---

[1] Virgil, " Æn.," iii. 426.   [2] Virgil, " Æn.," vii. 674.
[3] Lecky's " Hiftory of European Morals," vol. i., p. 165.

mock at the popular gods, and naturally at fuch mythological monfters and heroes as Homer had reverently recounted as feen by his hero in Hades, Tityos lying on nine roods of ground, and ever devoured by two vultures; Tantalus up to his chin in water, with the fineft fruit hanging before him from branches which he could never grafp, and the like.[1] The forms of thefe favourites of the poets lingered, however, in art; fculptors, painters, potters, glyptic artifts, gladly availed themfelves of their fantaftic fhapes, as had the old poets before them. Cyclops and the Harpies, Medufa's head and the hundred-eyed Argus are examples in point. Thus Pegafus becomes the type of Corinth on the coins of Auguftus, and the Sphinx of Egypt. The Siren, half-bird, half-virgin, reprefents Cumæ. The Chimæra is another emblem of Corinth. The Centaur Chiron and the Griffin are found on late coins dedicated to Apollo. Others are to be feen on bas-reliefs and vafes. If they ever poffeffed any conftraining moral or religious force, it has long evaporated; but the poet and the artift are ftill thankful for thefe old mythological forms. For them,

> "Vinctus fedet immanis ferpentibus Otos,
> Devinctum mæftus procul afpiciens Ephialten;"

and

> "Cerberus et diris flagrat latratibus ora,
> Anguibus hinc atque hinc horrent cui colla reflexis,
> Sanguineique micant ardorem luminis orbes."[2]

---

[1] "Od.," xi. 577 *feq.*
[2] Virgil, "Culex," 219, 233.

In the fame manner Spenfer depicts in a famous
ftanza the fingular group of objects drawn from
ancient monftrofities and romantic conceptions
with which were decorated the walls of the houfe
of Imagination :

> " His chamber was difpainted all within
>   With fondry colours, in the which were writ
>   Infinite fhapes of thinges difperfed thin ;
>   Some fuch as in the world were never yit,
>   Ne can devized be of mortall wit ;
>   Some daily feene and knowen by their names
>   Such as in idle fantafies do flit ;
>   Infernall hags, centaurs, feendes, hippodames,
> Apes, lyons, ægles, owles, fooles, lovers, children, dames."[1]

It can hardly be faid, therefore, as Aulus Gellius
too confidently affirms, that marvels and prodigies
fuch as we have named are of no importance,
" ad ornandum juvandumque ufum vitæ."[2]   He
himfelf, on landing from Greece at Brundifium,
tells us how eagerly he rufhed to a bookfeller's
fhop, bought up a quantity of books containing
fuch recitals at a cheap rate, and then devoured
them in two confecutive nights.   In fhort, the
imagination muft be fed, like the bodily appetite ;
and ftories of marvels muft at times be ferved up
to it, when they are as grateful after a period of
abftention as highly-feafoned viands are at certain
times to the bodily tafte.   They ferved for ruder
ages the fame end which our own novels of
character and flight incident perform for more
critical readers.   They amufe and infenfibly inftruct.
It was impoffible for an ancient Greek to liften to

[1] " Faeric Queene," ii. 9, 50.
[2] Aulus Gellius, ix. 4.

the hunting of the mighty Calydonian boar without his own pulfes beating the quicker the next time he found himfelf chafing a dangerous quarry in the Theffalian mountains ; nor could he ever hear the recital of Cyclops's cannibal feaft and portentous gluttony, as told in the "Odyffey," without having his own character directed to that moderation and chaftened fpirit which are among the fpecial attributes of his nation. Though difcredited, they ftill hold their own in the national Olympus, mutely inculcating a horror of the monftrous appetites of favagery.

He who would gauge the credulity of our fore-fathers in the matter of monfters fhould confult Topfel's "Hiftory of Four-footed Beafts," 1658. There he will find marvellous accounts and illuf-trations of the fea-horfe, the fu, the water-fheep, the tartarine, and the mantichora. Topfel ob-tained his notion of this horrific creature from Ctefias, but his print of it is fo amazing, that it was certainly evolved from imagination.

CHAPTER XVII.

OYSTERS AND PEARLS.

"Parum fcilicet fuerat in gulas condi maria, nifi manibus, auribus, capite, totoque corpore a fœminis juxta virifque geftarentur."—(PLINY, *Nat. Hift.*, ix. 35.)

HIGHLY prized as pearls have been whenever they could be procured, the Greeks feem to have known little or nothing of them ; and yet the Phœnicians, thofe mafter-mariners of antiquity, might well be fuppofed to have trafficked in them, when they

> "Saw the merry Grecian coafter come,
>   Freighted with amber grapes, and Chian wine,
>   Green burfting figs, and tunnies fteeped in brine,
> And knew the intruders on their ancient home."

It might have been thought, too, that Homer would have hung a carcanet of pearls round Helen's neck, or powdered the braided treffes of Circe and Calypfo with them, when he wifhed to enhance their beauty. Until the firft century before Chrift they were not abundant, or objects of ordinary luxury at Rome. During the reigns of the Cæfars, in the firft century after Chrift, pearls were highly valued, and were prominently difplayed by the Romans at

> " Their fumptuous gluttonies and gorgeous feafts
> On citron tables or Atlantic ftone,
> Their wines of Setia, Cales and Falerne,
> Chios, and Crete,"

when they would

> " Quaff in gold,
> Cryftal, and myrrhine cups, emboffed with gems
> And ftuds of pearl."[1]

The oyfter, however, was well known to the Greeks. In early times, indeed, it feems to have been curioufly defpifed as an article of food. The only time that it is mentioned in Homer is when Patroclus, in the " Iliad," hurls Cebriones, the charioteer of Hector, from his place in the chariot, and, after the fafhion of the time, mocks him : " Ye gods! truly he is an active man! How cleverly he dives! If, indeed, he were on the fifhy fea, this fellow would fatisfy many men by groping on the bottom for oyfters, leaping off his fhip even if it were very ftormy weather, fo cleverly does he now dive head-foremoft from his chariot to the plain!"[2] This paffage is curious both in itfelf, and alfo becaufe it was much ufed in controverfy by the Chorizontes (thofe who would affign the " Iliad " and " Odyffey " to different authors), inafmuch as the Homer of the " Iliad," it was faid, does not introduce his heroes as eaters

[1] "Par. Regained," iv. 114.
[2] "Il.," xvi. 145.

of fifh, but the author of the "Odyffey" does
("Odyffey," xii. 330-332). On the other hand, it
was replied by the Scholiaft with a delightfully un-
fcientific, if conclufive, argument, that they who
are accuftomed to eat oyfters may be confidered to
know the ufe of fifh in diet. A Homeric hero,
however, would as foon have thought of eating
fifh as a hero of Dr. Johnfon's time would have
drunk claret.[1]   Ariftotle gives an elaborate account
of the oyfter's habits and anatomy : " It has the
ftrangeft nature of all creatures, its body being
altogether concealed in fhell. It poffeffes two
openings, fome little diftance from each other,
very fmall, and not eafy to be difcerned, by
means of which it takes in and fends out water ;"
and more of fimilar import.   He treats its fenfes
with fcant reverence ; but we know that an oyfter
poffeffes heart, liver, mouth, gills, and other
organs, to fay nothing of a capacious ftomach and
ciliary appendages, which bring a conftant ftream
of water and food to its mouth.   What chiefly
ftruck the ancient Greeks with regard to its
economy is what firft impreffes a child at prefent,
the clofe manner in which it clings to the rock.
Plato employs this habit of the oyfter in a beautiful
paffage : " We have given a true account of the
foul," he fays, " in its prefent appearance ; but we
have looked at it in a ftate like that of the fea-god
Glaucus, whofe original nature can no longer be

---

[1] "Sir, claret is the liquor for boys ; port for men : but he
who afpires to be a hero" (fmiling) "muft drink brandy."
(Bofwell, vol. iii., p. 411, ed. 1816.)

readily diſcerned by the eye, becauſe the old members of his body have been either broken off or cruſhed, and in every way marred by the action of the waves; and becauſe extraneous ſubſtances—like oyſters, and ſea-weeds, and ſtones—have attached themſelves to him, ſo that he reſembles any other monſter than his natural ſhape: ſo with reſpect to the ſoul, we behold it affected by ten thouſand evils." And he continues: "We muſt look at its philoſophical nature, and muſt conſider to what it clings and what company it longs for, inaſmuch as it is kindred with the Divine, and the Immortal, and the Ever-exiſting; and what it would become were it wholly to follow theſe attributes, and by this impulſe be borne upwards out of the ſea in which it now lies, and diſencumbered of the ſtones and oyſters, and the many earthy, ſtony, and harſh ſubſtances, which have clung to it in conſequence of its feaſting upon earth, at thoſe banquetings which are deemed ſo happy."[1]

Pliny diſregards the oyſter in compariſon with its paraſite, the pearl. It furniſhes him with ſorrowful reflections upon the luxury of his age, the coſtlineſs and hazard with which it is ſought for: "Principium culmenque omnium rerum pretii margaritæ tenent." Fine pearls are ſupplied by the Indian Ocean; "and yet, to come by them, we muſt go and ſearch among thoſe huge and terrible monſters of the ſea which we have ſpoken of before. We muſt paſs over ſo many

[1] Plato, "Repub.," 611 D.

feas, and faile into far countries fo remote, and
come into thofe parts where the heate of the fun is
fo exceffive and extreme, and, when all is done,
we may perhaps miffe of them." But the beft
are found in the Perfian Gulf. Profeffor Skeat
confiders the word " pearl " derived from the Low
Latin " perula " or " pirula," a little pear, the
diminutive of " pirum." Whether from defign or
mifprint, his view is curioufly borne out by
Holland in the following words : " This fhell-fifh,
which is the mother of Pearle, differs not much
in the maner of breeding and generation from the
Oyfters; for when the feafon of the yeare requireth
that they fhould engender, they feeme to yawne
and gape, and fo do open wide; and then (by
report) they conceive a certaine moift dewe as
feed, wherewith they fwell and grow big, and
when time commeth labor to be delivered thereof;
and the fruit of thefe fhell-fifhes are the Peares
[*fic*], better or worfe, great or fmall, according to
the qualitie and quantitie of the dew which they
received. For if the dew were pure and cleare
which went into them, then are the Pearles white,
faire, and Orient; but if groffe and troubled, the
Pearles likewife are dimme, foul, and dufkifh."
This conceit of pearls being fprung from dew runs
through much mediæval poetry, and is a favourite
fancy with theologians. What was regarded as
playful imagination in Lord Beaconsfield's ftory of
the jeweller coming down once a year to wipe the
duchefs's pearls and lay them gently in the fun
with a fouth wind, has its prototype in Pliny.

He ſays the colour of pearls becomes yellow or remains white, like the complexion, according as they are expoſed to much or little ſunſhine. And yet, " as orient as they be, they waxe yellow with age, become riveled, and looke dead, without any lively vigor; ſo as that commendable orient luſtre '(ſo much ſought for of our great lords and coſtly dames) continueth but in their youth and decaieth with yeares."

Some ſhells were kept at Rome for perfume-caſes, and in them the pearls were left adhering to the halves. The pearl itſelf was ſuppoſed by Pliny to be ſoft and tender in the water, but to grow hard when once removed. If he exaggerates the danger of the ſhell cloſing upon the hand, he does not ſufficiently dwell upon the perils which the ſhell-divers run from the attacks of ſharks. Theſe are their moſt dreaded foes. It is worth while tranſcribing ſome more of his quaint fancies in Holland's words : " Let the fiſher looke well to his fingers, for if ſhe catch his hand between, off it goeth; ſo trenchant and ſharp an edge ſhe carrieth, that is able to cut it quite a-two. And verily this is a juſt puniſhment for the ˋtheefe, and none more: albeit ſhe be furniſhed and armed with other means of revenge. For they keep for the moſt part about craggie rocks, and are there found; and if they lie in the deepe, accompanied lightly they are with curſt ſea-dogs. And yet all this will not ſerve to ſkar men away from fiſhing after them; for why? our dames and gentlewomen muſt have their eares behanged

with them, there is no remedie. Some say that
these mother-pearles have their kings and captaines,
as Bees have; that as they have their swarmes led
by a master-Bee, so every troup and companie of
these have one speciall great and old one to con-
duct it, and such commonly have a singular dexteritie,
and wonderfull gift to prevent and avoid all
daungers. These they be that the dyvers after
pearles are most carefull to come by, for if they be
once caught, the rest scatter asunder and be soone
taken up within the nets." He knew of their
being laid in heaps, as at present, until, on the
creature dying, the pearls are found in the shells.
A good pearl ought to possess five qualities: it
should be orient (glittering), white, great, round,
smooth, and weighty. The best pearls, when these
qualities meet in them, were known at Rome as
" Unions," " as a man would say Singular, and by
themselves alone. The Greeks have no such
tearmes for them, neither know they how to cal
them; nor yet the Barbarians, who found them
first out, otherwise than Margaritæ."[1] Their
highest praise, he adds, is to be called *exaluminati*,
i.e. orient, and clear as alum. Pear-shaped pearls,
(or *elenchi*) were greatly valued ; " our dames take
a great pride in a brauerie, to haue these not only
hang dangling at their fingers, but also two or three
of them together pendant at their eares." They took
pleasure in hearing them when thus hung knock

[1] This word " margarita," so well-known in modern languages,
is said to be derived from a Sanscrit word manâarîtâ, " the
pure." (See Trench, " Parables," 6th ed., p. 130.)

together like cymbals; hence ſuch pearls were called *crotalia.* All this luxury once more tempts him to moralize. " Now adayes alſo it is growne to this paſſe, that meane women and poore mens wives affeⱺ to weare them, becauſe they would be thought rich, and a by-word it is amongſt them, That a faire pearle at a woman's eare is as good in the ſtreet where ſhe goeth as an huiſher to make way, for that every one will give ſuch the place.[1] Nay, our gentlewomen are come now to weare them upon their feet, and not at their ſhoo-latchets only, but alſo vpon their ſtartups and fine buſkins, which they garniſh all ouer with pearle. For it will not ſuffice nor ſerue their turne to carie pearles about them, but they muſt tread upon pearles, goe among pearles, and walke, as it were, on a pauement of pearle."

Some pearles, but few and ſmall ones, were found in the Boſphorus, and off the coaſts of Acarnania and Mauretania. From theſe Pliny paſſes to a notorious example of waſteful exceſs and intolerable pride. " I myſelfe have ſeen Lollia Paulina (late wife and after widdow to Caius Caligula the emperor), when ſhe was dreſſed and ſet out, not in ſtately wiſe nor of purpoſe for ſome great ſolemnity, but only when ſhe was to go to a wedding ſupper, or rather unto a feaſt

---

[1] To ſhow how diffuſe is Dr. Philemon Holland, it is worth while contraſting this ſentence with the terſe beauty of the original : "Affeⱺantque jam et pauperes, liⱺorem fœminæ in publico unionem eſſe diⱺitantes." Yet is Holland's quaintneſs not diſpleaſing. " Huiſher " is our modern " uſher," and " ſtart-ups " are high ſhoes.

when the aſſurance was made, and great perſons they were not that made the ſaid feaſt," (mediocrium etiam ſponſalium cœna),—" I have ſeen her, I ſay, ſo befet and bedeckt all over with hemeraulds and pearles, diſpoſed in rewes, ranks, and courſes, one by another, round about the attire of her head, her cawle, her borders, her peruk of hair, her bondgrace" (high hood over the forehead), "and chaplet; at her ears pendant, about her neck in a carcanet, upon her wreſt in bracelets, and on her fingers in rings, that ſhe glittered and ſhon again like the ſun as ſhe went. The value of theſe ornaments ſhe eſteemed and rated at 400 hundred thouſand Seſtertii." When Pliny mentally compares this luxury with the ſimple array of ſuch old Romans as Curius or Fabricius even while triumphing, he cannot forbear bitter reflections. Who would not have wiſhed that they had been pulled out of their chariots and never triumphed, than that their victories ſhould have let into Rome ſuch a flood of coſtly ornaments! As for Lollia herſelf, ſhe may fitly point a moral. All theſe jewels came to her from her uncle, M. Lollius, and were the fruit of his extortions and outrageous exactions from different provinces. Yet the end was that, loſing the friendſhip of Caligula, and being accuſed of bribery and corruption, he "dranke a cup of poiſon, and preuented his judiciall trial; that forſooth his neece Lollia, all to be hanged with jewels of 400 hundred thouſand Seſtertii, ſhould be ſeene glittering and looked at of euery man by candle-light all a ſupper time."

Pliny alfo tells the ftory of Cleopatra and the precious pearl, which fhe diffolved in vinegar and then fwallowed, in order to carry out her vainglorious boaft that her fupper fhould coft her fixty millions of fefterces. Clodius, however, the fon of a tragic poet, had long before her time performed the fame fenfelefs feat. After the taking of Alexandria, pearls were common at Rome. Arabia, Pliny notes, if bleffed in its perfumes, is ftill more enriched by its feas and the abundance of pearls which they produce.[1]

In the Old Teftament, the word " pearl" is fuppofed to mean " mother-of-pearl," or " cryftal," or "rubies." The pearl proper was not known to the Jews until later times; it often appears in the imagery of the New Teftament.

The true pearl-oyfter is the *avicula margaritifera* of the Perfian Gulf, Cape Comorin, and Ceylon ; but in Britain pearls are found in the *unio margaritiferus*, in the *oftrea edulis* (oyfter), and even in the *mytilus communis* (common muffel), though thefe are not fo valuable. The white iridefcent mother-of-pearl fubftance in thefe fhells is known as " nacre." It is compofed of layers of membranous fhell-fubftance. The pearl itfelf is merely an accretion of nacre, generally round fome fubftance of foreign origin which has found its way into the fhell. Hence artificial pearls have been procured by wounding the creature with a fharp-pointed implement or introducing foreign bodies.

[1] See Pliny (Holland's Tranflation), ix. 35 ; xii. 18 ; and Hor., ii. 4, 239.

The *ostrea edulis,* although pearls are found in it, was in Roman times, as in ours, far more celebrated at feasts. It may be said to have its capital in Britain (says Professor E. Forbes), although it is found elsewhere on the coasts of Europe. It has always been esteemed best from the beds off Kent. In Roman times, an epicure could distinguish the British oysters at once:

> " Circeis nata forent an
> Lucrinum ad faxum, Rutupinove edita fundo
> Oftrea, callebat primo deprendere morfu ;"[1]

just as anyone can at present tell a native from the huge coarse oyster of Cleethorpes.

The Romans knew of, and prized our British pearls. Indeed, Suetonius intimates that they formed the chief inducement to Cæsar to invade Britain. Pliny characterizes them fairly, as being small and poorly coloured; and he knew them well, as the breastplate which Julius Cæsar dedicated to Venus Genetrix at Rome was composed of them. Those rivers, with us, which flow from mountains generally contain the pearl-shells. The Esk and Conway are famous for them. A Conway pearl is said to be inserted in the royal crown of England. The Irt, in Cumberland, also produces pearls; but the most famous of our pearl-bearing rivers, in ancient as in modern times, was undoubtedly the Tay. We have examined many which were found in this river in recent years. They are all wanting in brilliancy—are not orient, in short. The best have a slightly pink tinge.

[1] Juv., iv. 140.

Tacitus, like Pliny, moralizes over pearls. To both writers they were the symbol of unbounded luxury. "The ocean round Britain produces pearls, but they are dusky and of a livid. hue. Some think that those who collect them are wanting in art, for, in the Red Sea, pearls are taken out from their shells while living and yet breathing; in Britain they are collected just as they have been expelled by the pearl-oyster. I would sooner believe that fine properties were wanting to the pearls than avarice in us."[1]

Among the gifts which Ovid feigns Pygmalion to have heaped on the statue of the nymph whom he loved, are gems for the fingers and necklaces for her slender neck:

"Aure leves baccæ, redimicula pectore pendent,
    Cuncta decent."

Virgil, too, when speaking of the blissful life of the shepherd, says,—what if he has none of the refinements of luxury :—

"Nec Indi
Conchea bacca maris pretio est ; at pectore puro
Sæpe super tenero prosternit gramine corpus."

Indeed, " bacca " or " berry," with some poetic addition, was a usual designation for a pearl. " Variis spirat Nereia bacca figuris," says Claudian ; sometimes by itself:

"Quin et Sidonias chlamydes, et cingula baccis
Aspera, gemmatasque togas
Dividis ex æquo."

---

[1] Tac. "Ag.," 12. British pearls with Pliny are " parvos atque decolores ;" with Tacitus, " subfusca ac liventia."

And ſtill more clearly, with ſome epithet :

> "Nec ſit marita quæ rotundioribus
> Onuſta baccis ambulet."[1]

Shakeſpeare does not ſeem to have been fond of pearls ; he loves, indeed, the "liquid pearl" on the "bladed graſs," but does not go out of his way to dwell upon the beauty or rarity of the ornament. He had read Pliny, however, as appears from what Troilus ſays of Creſſida :

> "Her bed is India ; there ſhe lies, a pearl."

With Milton, the pearl forms part, not only of his claſſical imagery, but alſo of his deep ſenſe of the beauty that dwells in all harmonious and regular ſights and ſounds. He knew the mediæval conceit of tears changing into pearls :

> "The fair bloſſom hangs the head
> Sideways, as on a dying bed,
> And thoſe pearls of dew ſhe wears,
> Prove to be preſaging tears."[2]

With him, too,

> "Morn, her roſy ſteps in the eaſtern clime
> Advancing, ſowed the earth with orient pearl."[3]

And in Paradiſe,

> "From that ſapphire fount the criſped brooks,
> Rolling on orient pearl and ſands of gold,
> With mazy error, under pendent ſhades
> Ran nectar."

In heaven, too, there is a bright ſea "of jaſper, or

---

[1] Ovid, "Met.," x. 265 ; Virgil, "Cul.," 67 ; Claud., "Cons. Honor," 592 ; and "Laud. Stil.," ii. 88 ; Hor., "Ep.," viii. 13.
[2] "Epitaph on the Marchioneſs of Wincheſter."
[3] "Par. Loſt," v. i.

of liquid pearl;" while as for the pearls of actual daily life,

> "The gorgeous Eaſt, with richeſt hand,
> Show'rs on her kings barbaric pearl and gold."

And, faireſt ſcene of all, when Sabrina "commended her innocence" to the Severn's flood:

> "The water nymphs that in the bottom played,
> Held up their pearled wriſts and took her in."[1]

Who that has not ſeen them for himſelf, if he loves to muſe near running water, hold up their pearled wriſts as the long-ſwaying treſſes of the water ranunculus, with their white bloſſoms, riſe to the ſurface and again gracefully ſink?

Pearls are only one item in the long liſt of woman's adornments which ſo characteriſtically call forth the anger of Burton: "Why do they adorn themſelves with ſo many colours of pearls, fictitious flowers, curious needle-works, quaint devices, ſweet-ſmelling odours, with thoſe ineſtimable riches of precious ſtones, pearls, rubies, diamonds, emeralds? etc. Why do they crown themſelves with gold and ſilver, uſe coronets and tires of ſeveral faſhions, deck themſelves with pendants, bracelets, ear-rings, chains, girdles, rings, furs, ſpangles, embroyderies, ſhadows rebatoes, verſicolor ribbands? Why do they make ſuch glorious ſhows with their ſcarfs, feathers, fans, maſks, furs, laces, tiffanies, ruffs, falls, calls, cuffs, damaſks, velvets, tinſels, cloth of gold, ſilver tiſſue?" etc. And then the old miſogyniſt concludes:

[1] See "Par. Loſt," iv. 238; ii. 4; "Comus," 834.

"They had more need, ſome of them, be tied in bedlam with iron chains; have a whip for a fan, and hair-cloths next to their ſkins, inſtead of wrought ſmocks; and have their cheeks ſtigmatiſed with a hot iron, I ſay, ſome of our Jezebels, inſtead of painting, if they were well ſerved."

"Pars minima eſt ipſa puella ſui."[1]

---

[1] "Anatomy of Melancholy," Part iii. 2, 3.